Neues verkehrswissenschaftliches Journal

Ausgabe 19

Modeling of Various Mixed Traffic Zones for Evaluation of Operating Performance on Urban Rail-bound Networks

(Modellierung unterschiedlicher Mischverkehrszonen zur Bewertung des Leistungsverhaltens von Stadtbahnnetzen)

Von der Fakultät Bau- und Umweltingenieurwissenschaften

der Universität Stuttgart zur Erlangung der Würde einer

Doktor-Ingenieurin (Dr.-Ing.) genehmigte Abhandlung

Vorgelegt von

Di Liu

aus Heilongjiang, VR China

Hauptberichter: Prof. Dr.-Ing. Ullrich Martin

Mitberichter: Prof. Dr.-Ing. Nils Nießen

Tag der mündlichen Prüfung: 15.02.2017

Institut für Eisenbahn- und Verkehrswesen der Universität Stuttgart

Februar 2017

ISBN 9783744841030

Preface

Evaluating the operating performance is one of the most important goals during capacity research for the macroscopic and mesoscopic evaluation of railway systems. The relationship between the quality of operation and the capacity of the investigated network can only be determined with a given operating program and further optimized based on proper measures. Two basic methods used in capacity research for railway systems are the analytic method and the simulation method. The waiting time function is an important intermediate result for the derivation of the recommended area of traffic flow. Previous research on this subject was conducted in the early 90s by Hertel and Ludwig, and the simulation method was further developed by Schmidt and Chu for use in practical applications.

Previous capacity research has mainly focused on conventional (heavy) railway systems; little research has been done on urban rail-bound transport with consideration of road traffic influences. In urban mixed traffic zones, the traffic is a mixed system and in addition to the perturbations caused by the railway system itself, the external influences caused by road traffic also have to be considered. The interactions between the urban rail-bound transport and road traffic can lead to extra waiting time for the urban rail-bound transport and influence the quality of its operation or capacity. For the various types of urban mixed traffic zones, the corresponding road traffic influences depend on the behavior and structure of the various modes of road traffic that come in contact with the rail-bound transport.

One of the most important results of this research is the model built to evaluate the road traffic influences on urban rail-bound transport and to determine the overall waiting time with a given operating program in an investigated area with one or more mixed traffic zones. In addition, the simulation method for capacity research was used in modeling various urban mixed traffic zones to determine their operating performance.

This dissertation is based on the DFG-research project "Capacity Research in Urban Rail-Bound Transportation with Special Consideration of Mixed Traffic"; Ms. Liu was the major researcher of this project. In this dissertation, the DFG-research is expanded to also include various types of urban mixed traffic zones. The simulation-based approaches can be used for appropriate investigated cases to estimate the through-put capacity and the extra waiting time the urban rail-bound transport will incur due to the external influences of the road traffic. The algorithm developed as a part of this research is used for modeling various urban mixed traffic zones and can successfully evaluate the operating performance by determining the waiting time function with event-driven systems. Due to the complexity of the interactions among urban mixed traffic in urban mixed traffic zones, a new model function of the waiting time function was developed to more precisely incorporate the changes of the waiting time with the increase of traffic flow of urban rail-bound transport. Therefore due to this research, more plausible results can be obtained for the capacity research of urban rail-bound transport with road traffic influences in an investigated area with mixed traffic zones .

Stuttgart, March 2017

Ullrich Martin

Acknowledgement

I would like to thank those who supported and encouraged me during my graduate study over the past years. First of all, I would like to thank my parents. They inspired and motivated me to study abroad positively. Without their encouragement, it is impossible for me to accomplish my dissertation work.

I want to sincerely thank Prof. Dr.-Ing. Ullrich Martin who helped and guided me on my scientific research. He was very important for me during my years at University of Stuttgart and will be so for my career life in the future. I am particularly grateful to him for his strict requirements and valuable comments in the field of research and my dissertation. His constructive suggestions helped me conduct research more effectively. I also highly appreciate his guidance on the scientific and teaching works that expanded my knowledge base and improved my skills in the past few years.

I would like to express my sincerest appreciation to my dear colleges, who supported and helped me a lot not only in the field of research and institute works, but also in my daily life. I want to thank Dr.-Ing. Yong Cui who helped me to overcome the challenges while writing this dissertation.

Table of Content

Preface 5

Acknowledgement 7

Table of Content 9

List of Figures 13

List of Tables 17

Abstract 19

Zusammenfassung 21

1   Introduction 23

2   Basis for Mixed Traffic Influences on Urban Rail-bound Transport 27
2.1   Urban Mixed Traffic Influences 27
2.1.1   Urban Mixed Traffic at Urban Mixed Traffic Zone 27
2.1.2   Traffic Control Signaling System 30
2.1.3   Interactions among Urban Mixed Traffic 34
2.1.4   Special Conditions 37
2.2   Capacity Research for Urban Rail-bound Transport 39
2.2.1   Basic Terms 40
2.2.2   Methods of Capacity Research 43
2.2.3   Structure of Capacity Research 47
2.3   Essential Tasks of Capacity Research for Urban Mixed Traffic 52
2.4   Objectives of the Dissertation 53

3   Approaches for Description of Road Traffic Influences on Urban Rail-
         bound Transport 57
3.1   Overview 57
3.2   Modeling Approach 57
3.2.1   Basic Concept 57
3.2.2   Methodology of Modeling Approach 58
3.2.3   Capacity Research based on Modeling Approach 65

3.3   Distribution Approach 69

3.3.1   Basic Concept 69

3.3.2   Perturbation Parameters 71

3.3.3   Methodology of Distribution Approach 73

3.3.4   Capacity Research with Distribution Approach 78

3.4   Conclusion 81

4   Modeling of an Adapted Waiting Time Function 83

4.1   Overview 83

4.2   Existing Waiting Time Function 83

4.2.1   Waiting Time Function by Ludwig 84

4.2.2   Waiting Time Function by Schmidt and Chu 86

4.3   Terms of Adapted Waiting Time Function 89

4.3.1   Overview 90

4.3.2   Adapted waiting time function with a new constant term 91

4.3.3   Adapted waiting time function with new term related to traffic flow 94

4.4   Conclusion 98

5   Algorithm for Evaluation of Mixed Traffic Zone 101

5.1   Basic Concept 101

5.2   Algorithm Description 102

5.2.1   Overview 102

5.2.2   Input Settings 105

5.2.3   Relevant Rules of Traffic Control Signal 119

5.3   Event-driven Cases 125

5.3.1   Overview 126

5.3.2 LC Cases for Level Crossing 129

5.3.3 SR Cases 1 and 2 for Shared Road 144

5.3.4 SS Cases 1 and 2 for Shared Space 147

5.4   Calculation of Indicators 149

5.4.1   Indicators of Algorithm for Level Crossing 150

5.4.2   Indicators of Algorithm for Shared Road 153

5.4.3   Indicators of Algorithm for Shared Space 160

5.5   Determination of Results for Capacity Research 163

5.5.1 Examining a Single Mixed Traffic Zone 164

5.5.2 Examining the Whole Investigated Area 168

5.6 Conclusion 171

6 Summary and Future Development 172

Appendix I: Basic Information of the Investigated Example 175

Appendix II: Results of Capacity Research for the Investigated Example 177

Appendix III: Supplementary Logics of Execution of Cases for Level Crossing 183

List of Abbreviations 187

Formula Symbols 188

Glossary 192

Bibliography 196

List of Figures

Figure 2-1: Classification of Mixed Traffic 28

Figure 2-2: Classification of Mixed Traffic Zones...29

Figure 2-3: Structure of Capacity Research (Source: updated based on [Martin et al. 2012a])...48

Figure 2-4: Evaluation of Performance for Capacity Research (Source: modified based on [Li 2015] & [Martin & Liu 2016a]) 49

Figure 2-5: Relationship between Entry Traffic Flow and Exit Traffic Flow for Determination of Throughput Capacity (Source: modified based on [Chu 2014]).......50

Figure 2-6: General Tasks...54

Figure 3-1: Representation of the Established Infrastructure of Mixed Traffic Zone of Level Crossing in Simulation Model (Source: modified based on [Martin & Liu 2016a]) 60

Figure 3-2: Established Infrastructure of Shared Road in Simulation Model 63

Figure 3-3: Workflow of Capacity Research based on Modeling Approach...............66

Figure 3-4: Waiting Time Function and Recommended Area of Traffic Flow based on Modeling Approach, compared with that without Road Traffic Influences (Source: modified based on [Martin & Liu 2016a] & [Martin & Liu 2016b])..............................68

Figure 3-5: Procedure for the Determination of Perturbation Parameters P_{ROT} for the Running Time Extension or Departure Time Extension (Source: modified based on [Martin & Liu 2016a])...77

Figure 3-6: Workflow of Capacity Research based on Distribution Approach (Source: modified based on [Martin & Liu 2016a])...79

Figure 3-7: Waiting Time Function and Recommended Area of Traffic Flow based on Distribution Approach, compared with that without Road Traffic Influences (Source: modified based on [Martin & Liu 2016a] & [Martin & Liu 2016b])..............................80

Figure 4-1: Fitting Curve with the Model Function of Waiting Time Function (4-9) with a Constant Term with Parameter d_3 (Source: modified based on [Martin & Liu 2016a])... 93

Figure 4-2: Fitting Curve with the Model Function of Waiting Time Function (4-10) with an additional Term $d_4/(1-\eta)$ (Source: modified based on [Martin & Liu 2016a]) 95

Figure 4-3: Fitting Curve with the Model Function of Waiting Time Function (4-11) with an additional Term $d_5/(1-\eta)^{2/3}$ (Source: modified based on [Martin & Liu 2016a])... 97

Figure 5-1: Scheme of the Algorithm Realized in the Model of Event-driven System (Source: modified based on [Martin & Liu 2016a]).. 104

Figure 5-2: Microscopic Description of Routes of Urban Mixed Traffic at Mixed Traffic Zones (Source: developed based on [Martin & Liu 2016a])..................................... 109

Figure 5-3: Exclusive Routes and Compatible Routes (Source: modified based on [Martin & Liu 2016a]).. 112

Figure 5-4: The Effective Occupation Time of Road Traffic Route with Different Traffic Light Phases of Traffic Control Signal (Source: modified based on [Martin & Liu 2016a])... 115

Figure 5-5: Route Set of Road Traffic... 123

Figure 5-6: The Workflow of Event-Driven Simulation with the Developed Algorithm¯ (Source: modified based on [Martin & Liu 2016a]).. 127

Figure 5-7: Overview of the Algorithm for Various Mixed Traffic Zones.................. 129

Figure 5-8: Six LC Cases under Different Event Trigger Conditions for Mixed Traffic Zone of Level Crossing (Source: modified based on [Martin & Liu 2016a])............. 130

Figure 5-9: The Workflow of Execution of LC Case 1 134

Figure 5-10: The Probabilities for LC Case 3 137

Figure 5-11: The Workflow of Execution of LC Case 5 142

Figure 5-12: Workflow of Execution of LC Case 6.. 144

Figure 5-13: Workflow of SR Case for Mixed Traffic Zone of Shared Road 146

Figure 5-14: Workflow of SS Case for Mixed Traffic Zone of Shared Space 149

Figure 5-15: Relationship between Potential Occupation Time for Road Traffic (O_{ROT}) and the Hindrance Time of Urban Rail-bound Transport by Road Traffic (H_{URT}) (Source: [Martin & Liu 2016a] modified based on [Martin & Liu 2016b]) 166

List of Tables

Table 2-1: Special or Unusual Conditions 38

Table 4-1: Comparison of (Adjusted) Coefficient of Determination of the Adapted Waiting Time Function (4-9) and that of Existing Waiting Time Function (4-7) (Source: [Martin & Liu 2016a])..92

Table 4-2: Comparison of (Adjusted) Coefficient of Determination of the Waiting Time Function (4-10) and the Existing Waiting Time Function (4-7) (Source: [Martin & Liu 2016a]) 96

Table 4-3: Comparison of (Adjusted) Coefficient of Determination of the Adapted Waiting Time Functions with Four Parameters (4-10) and (4-11) (Source: [Martin & Liu 2016a]) 98

Table 5-1: The Locking Table of Route Related Locking for Urban Mixed Traffic (Source: modified based on [Martin & Liu 2016b] & [Martin & Liu 2016a])..............110

Table 5-2: Capacity Research Results of a Single Mixed Traffic Zone Using Various Methods 167

Table 5-3: Capacity Research Results of a Whole Investigated Area Using Various Methods 171

Abstract

The efficiency and operation quality of urban rail-bound transport are significantly influenced by the interfered road traffic in urban mixed traffic zones. The urban rail-bound transport is disturbed not only by perturbations in railway system, but also by the movements of road traffic. Therefore, it's very important to investigate the influences caused by road traffic in urban mixed traffic zone for capacity research on urban rail-bound transport. In this dissertation, the simulation method for capacity research is utilized to determine the road traffic influences on urban mixed traffic zones. The intermediate result of waiting time function is further developed to adapt to the investigated area with urban mixed traffic zone(s), which can further more precisely derive the recommended area of traffic flow.

Based on the works of capacity research for railway system by [Hertel 1992], [Schmidt 2009] and [Chu 2014], modeling of the road traffic influences on urban rail-bound transport at various types of urban mixed traffic zones is the focus of this dissertation. Furthermore, the extra waiting time of urban rail-bound transport caused by road traffic is determined by deriving the adapted waiting time function and recommended area of traffic flow to evaluate operating performance. The important new findings are listed below:

- According to the analysis of the interactions between road traffic and urban rail-bound transport in various urban mixed traffic zones, two approaches, modeling approach and distribution approach, are developed to model the road traffic influences. According to various investigated mixed traffic zones and available database, operating performance of urban rail-bound transport can be evaluated with either modeling approach or distribution approach.

- A newly developed model function for waiting time function is developed based on [Chu 2014] for adaption to the urban rail-bound system with consideration of external influences. It can represent the discrete points (waiting time of urban rail-bound transport with stepwise-varied traffic flows) better in the fitting curve. The (adjusted) coefficient of determination is higher than that with the existing waiting time function. Furthermore, the derived recommended area of traffic is more plausible.

- An algorithm for the determination of waiting time function and throughput capacity is developed based on the interactions among urban mixed traffic in various mixed traffic zones. It is able to build models for various types of urban mixed traffic zones. An event-driven system is used in this algorithm for modeling the urban mixed traffic zones, which is also a preliminary assessment of the significances of road traffic influences on each mixed traffic zone in a whole investigated area.

This research was based on the DFG-research project "Capacity Research in Urban Rail-bound Transportation with Special Consideration of Mixed Traffic" [Martin & Liu 2016a]. The capacity research for macroscopic and mesoscopic evaluation of railway system could be expanded to the urban mixed traffic based on the findings of this work. Furthermore, the quality of operation and capacity of urban rail-bound transport can be derived and evaluated with consideration of road traffic influences.

Zusammenfassung

Die Effizienz und Betriebsqualität des städtischen schienengebundenen Verkehrs wird stark durch den in städtischen Mischverkehrszonen störenden Straßenverkehr beeinflusst. Der städtische schienengebundene Verkehr wird nicht nur durch Störungen im Bahnsystem, sondern auch durch Bewegungen des Straßenverkehrs gestört. Es ist daher notwendig, Leistungsuntersuchungen im städtischen schienengebundenen Verkehr unter besonderer Berücksichtigung des Mischverkehrseinflusses zu untersuchen. In der vorliegenden Arbeit wird die Simulationsmethode für Leistungsuntersuchungen zur Ermittlung der Individualverkehrseinflüsse auf städtische Mischverkehrszonen genutzt. Die Wartezeitfunktion wird weiterentwickelt, um den optimalen Leistungsbereich an Untersuchungsräume mit städtischen Mischverkehrszonen anzupassen.

Basierend auf den Arbeiten auf dem Gebiet der Leistungsuntersuchung von [Hertel 1992], [Schmidt 2009] und [Chu 2014], liegt der Fokus dieser Dissertation auf der Modellierung der Einflüsse des Straßenverkehrs auf den städtischen schienengebundenen Verkehr in verschiedenen städtischen Mischverkehrszonen. Darüber hinaus wird die zusätzliche Wartezeit des städtischen schienengebundenen Verkehrs, die durch den Straßenverkehr verursacht wird, bestimmt, indem die angepasste Wartezeitfunktion und der optimaler Leistungsbereich abgeleitet werden, um die Betriebsleistung zu bewerten. Die wichtigsten neu gewonnenen Erkenntnisse werden im Folgenden aufgeführt:

- Nach der Analyse der Interaktion zwischen Straßenverkehr und städtischen schienengebundenen Verkehr in verschiedenen städtischen Mischverkehrszonen werden zwei Ansätze, ein Modellierungsansatz und ein Verteilungsansatz entwickelt, um die Individualverkehrseinflüsse zu modellieren. Gemäß den verschiedenen untersuchten Mischverkehrszonen und der verfügbaren Datengrundlage kann das Leistungsverhalten des städtischen schienengebundenen Verkehrs mit dem Modellierungsansatz oder dem Verteilungsansatz bewertet werden.

- Eine neu entwickelte Modellfunktion für die Wartezeitfunktion wird auf Basis von [Chu 2014] zur Anpassung an das städtische schienengebundene System unter Berücksichtigung von externen Einflüssen entwickelt. Diese kann die diskreten Datenpunkte (Wartezeiten des städtischen schienengebundenen Verkehrs mit schrittweiser Verdichtungsstufe) besser approximieren. Das (korrigierte) Bestimmtheitsmaß ist höher als dasjenige der bestehenden Wartezeitfunktion. Darüber hinaus ist der abgeleitete optimale Leistungsbereich für den hier untersuchten Anwendungsfall plausibler.

- Zur Bestimmung der Wartezeitfunktion und der durchsatzbezogenen Leistungsfähigkeit wurde auf Basis der Interaktion mit städtischem Mischverkehr in verschiedenen Mischverkehrszonen ein Algorithmus entwickelt, mit dessen Hilfe Modelle für verschiedene städtische Mischverkehrszonen modelliert werden können. In diesem Algorithmus wird ein ereignisgesteuertes System für die Modellierung der städtischen Mischverkehrszonen eingesetzt, das auch eine vorläufige Einschätzung der Bedeutung von Individualverkehrseinflüssen auf die Mischverkehrszonen im gesamten Untersuchungsraum ermöglicht.

- In der vorliegenden Arbeit, die im Kontext des DFG-Forschungsprojekts "Leistungsuntersuchungen im städtischen schienengebundenen Verkehr unter besonderer Berücksichtigung des Mischverkehrseinflusses " [Martin & Liu 2016a] (*English: "Capacity Research in Urban Rail-bound Transportation with Special Consideration of Mixed Traffic"*) entstanden ist, konnte das Gebiet der Leistungsuntersuchung zur makroskopischen und mesoskopischen Bewertung des Eisenbahnsystems um den städtischen Mischverkehr erweitert werden. Darüber hinaus kann die Betriebsqualität und die Leistungsfähigkeit des städtischen schienengebundenen Verkehrs unter Berücksichtigung von Individualverkehrseinflüssen abgeleitet und bewertet werden.

1 Introduction

The rail-bound transport in urban area plays an important role in mobility of public transport. The rail-bound transport in urban area differs from the pure railway system because it is operated in an urban mixed traffic zone with other transport modes including private transports and also other public transports. Therefore, other road traffic can also influence the efficiency of the operations of urban rail-bound transport, in addition to the operational hindrance between trains in pure railway system. Moreover, the traffic loads of road traffic obviously rise with the increase of demands and the economic development, which restricts the further extension of the existing rail-bound infrastructure in urban area. Hence, it is significant and meaningful to utilize the urban rail-bound transport as efficiently as possible. Capacity research on urban rail-bound transport is considered and developed in this dissertation to evaluate the operating performance that represents the inverse relationship between the capacity (i.e. traffic flow of urban rail-bound transport) and the quality of operation (inversely proportional to the average waiting time of urban rail-bound transports) in the investigated area with urban mixed traffic zone(s).

Previously researches were mainly focused on the rail-bound transport in pure railway system or road traffic on urban roadway. Nagel and Schreckenberg proposed to use the one-dimensional cellular automata model to simulate freeway traffic [Nagel & Schreckenberg 1992]. Other works were mainly focused on private car movements (see [Fukui & Ishibashi 1996] and [Newell 2002]) and pedestrian behaviors (see [Blue et al. 1997]). On the other hand, capacity research on railway system was developed step by step. It was firstly proposed by [Hertel et al. 1987] through analytical method and was further studied with simulation method for capacity research (see [Schmidt 2009] and [Chu 2014]).

However, it is barely that the mixed conditions are taken into consideration previously. In urban mixed traffic zones, the urban rail-bound transport is definitely impacted by the road traffic. In some cases, the urban rail-bound transport is significantly affected. In capacity research of urban mixed traffic zone, it cannot be ignored such impacts on the operation of urban rail-bound transport. Road traffic influences lead to increase of the waiting time of urban rail-bound transport, which further results in a deviation of the existing waiting time function. Therefore, the results of capacity research without taking into consideration of road traffic influences are inaccurate.

In this dissertation, further developments based on the findings of the DFG-research project [Martin & Liu 2016a] are studied. The mixed traffic zones are further widely categorized with shared space for pedestrians. Furthermore, a developed algorithm is expanded and further improved based on the basic theoretical algorithm in the DFG-research project [Martin & Liu 2016a]. The algorithm models for specific mixed traffic zones are established to determine and evaluate the operating performance of the urban rail-bound transport (URT) with consideration of road traffic (ROT) influences on the investigated area with mixed traffic zone(s).

In Chapter 2, the urban mixed traffic and urban mixed traffic zones that are the main research subjects in this dissertation are defined and discussed. Moreover, the interactions among the urban mixed traffic in different investigated mixed traffic zones will be described under various basic conditions. Afterwards, the existing findings and practical applications on capacity research with two main methods are generally introduced as a basis for capacity research in this dissertation.

Two developed approaches (modeling and distribution approach) developed in the DFG-research project [Martin & Liu 2016a], will be described in Chapter 3. The new improvement will be studied for modeling of the road traffic influences on urban rail-bound transport in various investigated mixed traffic zones, which is integrated with the existing simulation method for capacity research on urban mixed traffic zone.

Because of the external influences of road traffic on the mixed traffic zone, the existing waiting time function cannot be used to estimate the waiting time with consideration of road traffic influences properly. The deviation between the data points and the fitting curve will further affect the evaluation results. Therefore, in Chapter 4, new waiting time function proposed in the DFG-research project [Martin & Liu 2016a] considering the impacts on urban rail-bound transport from road traffic will be depicted.

In Chapter 5, the developed algorithm is studied on modeling event-driven system to evaluate various mixed traffic zones on the basis of the DFG-research project [Martin & Liu 2016a]. It is through statistics of random interactions among urban mixed traffic in the investigated mixed traffic zone. The evaluation of operating performance of the urban rail-bound transport with the road traffic influences in mixed traffic zone through determination of capacity research results with simulation method can be implemented. Finally, scientific results summaries and the future development will be discussed in Chapter 6.

2 Basis for Mixed Traffic Influences on Urban Rail-bound Transport

Capacity and operation quality of urban rail-bound transport at mixed traffic zones is certainly impacted by the mixed traffic. In order to carry out capacity research for urban rail-bound transport, the urban mixed traffic and the urban mixed traffic zone have to be comprehended as prerequisite. Definitions of urban mixed traffic in urban mixed traffic zones were preliminary presented in [Martin & Liu 2016a], which will be further discussed in Subchapter 2.1.1. Moreover, Subchapters 2.1.2 and 2.1.3 will describe the interactions among of urban mixed traffic in terms of conditions related to the traffic light control signaling system and discuss the research coverage in this dissertation. The relevant basic terms and current methods for capacity research are introduced in Subchapter 2.2. The main tasks of capacity research on urban rail-bound transport with consideration of mixed traffic zone and the objectives of this dissertation are clarified in Subchapters 2.3 and 2.4.

2.1 Urban Mixed Traffic Influences

2.1.1 Urban Mixed Traffic at Urban Mixed Traffic Zone

The urban mixed traffic is the general term for various modes of urban transports operated in urban mixed traffic zones. The urban mixed traffic includes urban rail-bound transport and other mutually interfered traffic in urban mixed traffic zone, which can be categorized into public transport (including urban rail-bound transport and buses), and private transport (including individual vehicles and pedestrians); or urban rail-bound transport and road traffic.

In order to evaluate the operating performance of urban rail-bound transport with consideration of the perturbations as a result of urban rail-bound transport themselves and the external influences of other modes of transports, as described in the DFG-research project [Martin & Liu 2016a] and shown in Figure 2-1, the basic categories in this dissertation are:

- Urban rail-bound transport
- Road traffic

The urban rail-bound transport includes all urban rail-bound transports operating with rolling stock on rail tracks of urban railway system, such as light rail transit (LRT) also known as "Stadtbahn" in Germany, and trams (streetcar) described as "Straßenbahn" in Germany. The road traffic can be further categorized into different groups:

- Motorized road traffic
- Non-motorized road traffic

Motorized road traffic includes various types of individual vehicles such as private cars, heavy vehicles (trucks), motor (-bicycles), and public street vehicles like buses. Non-motorized road traffic includes pedestrians and bicycles [Martin & Liu 2016a].

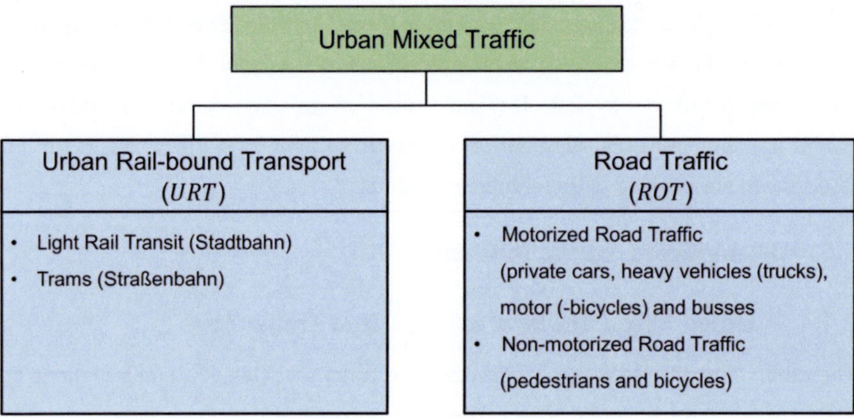

Figure 2-1: Classification of Mixed Traffic

Urban mixed traffic zone is the urban traffic area, where various interfered urban mixed traffic share the same road. The mixed traffic zone is diverse and complicate because of diversified topology structures and interactions among various urban mixed traffic. In the DFG-research project [Martin & Liu 2016a] basically divided urban mixed traffic zones into the following two categories:

- Level crossing (LC)
- Shared road (SR)

Additionally, another type of urban mixed traffic zone with rail tracks for urban rail-bound transport and pedestrians is discussed in this dissertation:

- Shared space (SS)

The shared space is an urban area that is planned to be shared by pedestrians and vehicles. Usually, the relevant facilities such as traffic control signals are removed.

There are different shared spaces for various shared motilities. The shared space for pedestrians and urban rail-bound transport (tram in general) with and without motorized road traffic (cars and buses) are discussed in this dissertation.

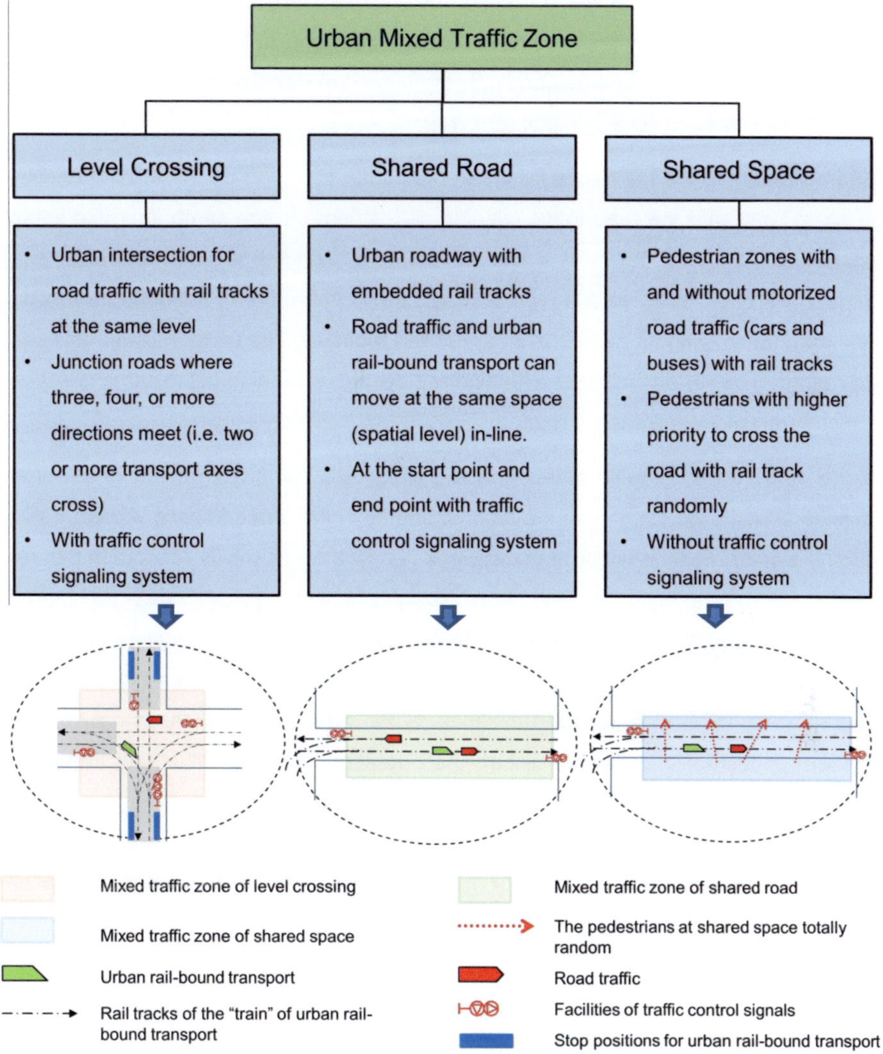

Figure 2-2: Classification of Mixed Traffic Zones

As a major type of urban mixed traffic zone discussed in this dissertation, as shown in Figure 2-2, level crossing is an urban intersection where an urban rail-bound vehicle can move on its rail track crossing a road or path at the same level [Martin & Liu

2016b]. The term also applies when an urban rail-bound line with separate right-of-way or reserved track crosses a road in the same manner. Level crossing also has some different categories, including crossing roads with four-direction cross and T-junction roads with three-direction cross etc.. Those categories are defined by the specific topological structures of urban mixed traffic zone of level crossings that are usually controlled by the correspondingly traffic control signaling system at the investigated level crossing (see Subchapter 2.1.2).

Shared road has the rail tracks for urban rail-bound transport on the urban roadway to share road with the road traffic (without pedestrians) at the same time and same space (spatial level). Basically, a shared road is a segment of urban roadways embedded with rail tracks. In this type of mixed traffic zone, urban rail-bound transport and road traffic (without pedestrians) are mixed together. The urban mixed traffic can be regarded as various vehicles with different dynamics running in the same direction concurrently along the shared road.

In the mixed traffic zone of shared space, pedestrians can cross rail tracks any time to pass through the street with the highest priority in the shared space without traffic control signals. The behavior of pedestrians is assumed as totally random in this research. Some pedestrians may stop crossing when they an urban rail-bound transport is approaching. Some pedestrians may stop crossing only when the urban rail-bound transport already approached. Some pedestrians may never stop regardless of urban rail-bound transport distance. When there is no motorized road traffic, random pedestrians' behaviors disturb the operation of urban rail-bound transport. Another possibility is that, some cars and buses are allowed to move or stop temporarily along the shared space. Due to the design principle for pedestrians inclusion, the traffic flow and speed limitation for motorized road traffic are both relatively low.

2.1.2   Traffic Control Signaling System

Traffic control signaling system controls and coordinates to ensure the movements of road traffic as smooth and safe as necessary. Various traffic control signaling systems are utilized to minimize the delays of traffic and to improve the utilization of road. In mixed traffic zone, the traffic control signaling system plays an important role particularly for level crossing and shared road. The traffic control signals are commonly installed at most urban intersections, some of which are mixed traffic zones (i.e. level crossings) with urban rail-bound transport.

There are various situations and settings of traffic control signals used to control and coordinate the movements of road traffic and the operation of urban rail-bound transport in different mixed traffic zones. In the DFG-research project [Martin & Liu 2016a], three major situations for mixed traffic zones were summarized. Those situations are respectively discussed below:

With traffic control signal

The simple traffic controller is the fixed time control (simple clockwork mechanisms), which coordinates the intersections based on scheduled time plan with normal rotation of traffic light phase in each traffic cycle. Further development of traffic control signaling system is self-adjusted system with computerized control and coordination. Sensors are installed as detectors, which are utilized to capture the present state of vehicles or road condition. The vehicle-dependent traffic control signaling system can adjust the corresponding traffic light cycles based on the actual traffic condition.

Most level crossings are controlled by the traffic control signals or the railroad crossing signals. The urban rail-bound transport can apply for the priority to pass through the level crossing. The application is sent by detectors along the rail track so that the information can be transmitted to the corresponding traffic control signal.

In general, the urban rail-bound transport has a higher priority than road traffic at level crossings. However, in reality, the urban rail-bound transport cannot have absolute priority all the time. There are regulations for road traffic to ensure the smooth and safe operation of urban mixed traffic in urban mixed traffic zones. As described in the DFG-research project [Martin & Liu 2016a], two main regulations of traffic control signaling system for road traffic are discussed in this dissertation, limitation of maximum red light phase and limitation of minimum green light phase, which suggested that the urban rail-bound transport cannot have the highest priority all the time.

The limitation of red light phase refers to the maximum time duration of red light phase for road traffic at level crossing. The regulated limitation is the time of red light phase for road traffic that cannot wait for its green light phase over the pre-defined maximum time duration. If the limitation of maximum red light phase for road traffic was reached, the urban rail-bound transport would lose its right of priority to pass through the level crossing. The road traffic exceeding the limitation of maximum red light phase can get higher priority than the urban rail-bound transport. Therefore, the urban rail-bound transport is influenced by the road traffic through the operation of the traffic control signal essentially.

In addition, the minimum green light phase is the minimum time duration of green traffic light phase of the traffic control signal for operation of road traffic, which can restrict the operation of urban rail-bound transport with higher priority at level crossing. The limitation of minimum green light phase regulates the time duration that road traffic have to complete under the control of traffic control signal with green traffic light. Even if an urban rail-bound transport asks for permission to proceed to pass through the level crossing, the road traffic directed by traffic control signals would yield right away if the road traffic already completes the minimum green light phase.

According to those basic regulations, it is possible that road traffic can cause external influences on urban rail-bound transport at the level crossing with relative complex structure and higher traffic flow of urban rail-bound transport. Urban rail-bound transports compete for the priority to ask traffic control signals for passing the level crossing not only with road traffic but also with each other. Road traffic will lose portion of their green light phase because of the higher priority of urban rail-bound transports before reaching the regulated limitations. Integrated with the regulation of limitation of minimum green light phase, road traffic can increase theirs priority. Then, the urban rail-bound transport will lose priority before road traffic completing minimum green light phase, which will lead to extra waiting time.

Theoretically, urban rail-bound transport can be affected by traffic control signals, even if there are very low traffic loads of road traffic or no road traffic at the level crossing [Martin & Liu 2016a]. However, such impacts are relatively mild in the operation process.

For level crossing, when no urban rail-bound transport arrives based on the scheduled timetable (or with some delays caused by perturbations), the road traffic continues on the scheduled traffic cycle (time) or self-adjusted system of traffic control signal. The road traffic influences when an urban rail-bound transport arrives and applies to pass through the level crossing. It is possible that a green light phase for road traffic is on progress. If this ongoing road traffic under green light phase conflicts with the coming urban rail-bound transport, the ongoing road traffic can finish its minimum green light phase based on the pre-defined regulation of traffic control signal. Here, the urban rail-bound transport experiences some waiting time as a result of road traffic influences. Afterwards, the urban rail-bound transport gets its permission to pass through the level crossing.

The urban mixed traffic along shared road is also controlled by traffic control signals[1]. Due to the interactions of urban mixed traffic as described in Subchapter 2.1.3, the urban rail-bound transport can be hindered by slow road traffic or the fulfilled shared road in front in the same direction. Since the urban roadway with rail tracks is shared by both urban rail-bound transport and road traffic, the mixed traffic zone of shared road is relatively small with low road traffic loads following the basic principle of practical condition for urban planning [Martin & Liu 2016a]. If the shared road has low traffic loads or even no road traffic during the low traffic flow period, there will be no traffic congestion along the shared road. The probability of road traffic influences on urban rail-bound transport is very low, which can be negligible during the investigation.

Without traffic control signal

For mixed traffic zone of level crossing without traffic control signals, the urban rail-bound transport can have the highest priority to pass through the level crossing. Generally, the level crossing is relatively small and the urban rail-bound transport in such mixed traffic zones is barely impacted by road traffic influences [Martin & Liu 2016a]. For safety consideration, there are no heavy traffic loads of road traffic or complex infrastructure for urban rail-bound transport.

On the other hand, the mixed traffic zone of shared space, normally, is designed without installation of traffic control signals. The driver of urban rail-bound transport has to pay attention to pedestrians crossing rail tracks casually, or cars and buses that stop temporarily in front of the train. Pedestrians can have the priority to cross the street, which may result in extra waiting time for urban rail-bound transport along the shared space.

---

[1] The traffic control signals here are normally installed at the next adjacent level crossing connected to the shared road, which can indirectly control the movements of urban mixed traffic along the shared road.

With railroad crossing signal

The traffic control signaling system with the railroad crossing signal usually consists of (flashing) red signal light[2] and a crossbuck for indication to road traffic. Under this situation, urban rail-bound transport has the absolute priority and can active the railroad crossing signal to indicate road traffic before the urban rail-bound transport arrives. The road traffic influences are approximately zero on urban rail-bound transport.

In conclusion, the road traffic influences on urban rail-bound transport will be investigated in this dissertation under three scenarios:

- Mixed traffic zone of level crossing with direct control of traffic control signal
- Mixed traffic zone of shared road with indirectly control of traffic control signal
- Mixed traffic zone of shared space without traffic control signal

2.1.3   Interactions among Urban Mixed Traffic

In order to evaluate the influences of road traffic on the urban rail-bound transport, it is necessary to consider the interactions between urban rail-bound transport and road traffic in various types of urban mixed traffic zones.

Based on summarized scenarios in Subchapter 2.1.2, the interactions can be reflected directly by the operations of corresponding traffic control signals at mixed traffic zone of level crossing. Since the urban roadways are shared with rail tracks, the urban rail-bound transport has to follow the movements and speed of road traffic in front in the mixed traffic zone of shared road. Meanwhile, the road traffic are under the control of traffic control signal at the next adjacent level crossing. Therefore, it is obvious that urban rail-bound transport and road traffic are controlled by the traffic control signaling system indirectly.

For the mixed traffic zone of level crossing and shared road, there can be different types of road traffic (private cars, heavy vehicles, bicycle and pedestrians) with diverse dynamics of vehicle movements, such as deceleration, acceleration and speeding. Even though the behaviors of various road traffic types are different, all

---

[2] Some traffic control signaling systems with the railroad crossing signal may also have yellow signal light before red signal light for indication of road traffic.

should follow the official regulations of traffic control signaling system at level crossings. In the mixed traffic zone of level crossing, the road traffic are directly controlled by traffic control signaling system, which furthermore influences the urban rail-bound transport. Therefore, private cars as the major type of road traffic primarily have some impacts on urban rail-bound transport. Besides, bicycles and pedestrians should also comply with the indication of traffic control signals for road traffic in mixed traffic zone of level crossing like private cars, regardless of whether they have specific lane or share the road with private cars.

Additionally, pedestrians can apply for the traffic control signals to pass through the level crossing. If there is an interaction between pedestrians and urban rail-bound transport on the road (level crossing), normally the urban rail-bound transport has the higher priority to pass. The pedestrians also have a limitation of maximum time duration for the red light phase. As a result, they don't need to wait too long for green light phase to pass through the crossing after application. For this situation, the application from pedestrians occurs much more randomly. The operation of urban rail-bound transport is based on the scheduled timetable, while the behavior of pedestrians is determined by the random target of trips, leading to random but relatively low possibility of pedestrian influences on the urban rail-bound transport.

Generally, both urban rail-bound transport and road traffic are stipulated by the officially defined speed limitations of the urban roadway. The movement of vehicle is under control of the corresponding speed limitations. Speeds must be adapted to the road traffic, visibility and weather conditions as well as the drivers' skills, characteristics of vehicles and traffic loads [Straßenverkehrs - Ordnung 2015]. For the various types of road traffic, the speed limitations vary as well. When different types of road traffic, such as bicycles and trucks exist, an accompanying change on road traffic speeds will be observed. For level crossing, the overall movement of road traffic is controlled by traffic control signals at level crossing. Road traffic influences are reflected on the operation of traffic control signal.

However, for mixed traffic zone of shared road, different road traffic influences are observed, due to the different dynamics of vehicle movements between urban rail-bound transport and road traffic. Road traffic affects the urban rail-bound transport, restricts the operation and slows down the speed of the urban rail-bound transport. Under this circumstance, the running time and waiting time of urban rail-bound transport are depend on the movements of road traffic in front. Therefore, the influences of different decelerations and accelerations of vehicle movements among urban mixed traffic reflected on the results are not significant. However, the impact of different speed limitations cannot be ignored.

Basically, there are two circumstances in which urban rail-bound transport may be hindered by road traffic. Urban rail-bound transport has to be move along with the road traffic at a relatively lower speed in front along the shared road. The other circumstance is that the road traffic with various types are congested in front of the urban rail-bound transport. In this situation, urban rail-bound transport is restricted by the movements of the congested road traffic. Ultimately, the waiting time is increased due to influences from road traffic that are controlled by traffic control signal at the next adjacent level crossing as described previously. Therefore, the speed limitations of different vehicle movements have to be considered, especially for road traffic with low speed that restricts urban rail-bound transport operation along the shared road.

For the mixed traffic zone of shared space, the situation is different, because of the lack of relevant facilities, such as road traffic control signals and road surface markings. The interactions between urban rail-bound transport and pedestrians highly depend on the random behaviors of pedestrians. Therefore, the influences of pedestrians cannot be ignored as those at level crossing. In shared space, the urban rail-bound transport is affected by the random behaviors of pedestrians or motorized road traffic with low traffic loads. If there are pedestrians crossing the rail tracks, the urban rail-bound transport has to brake to decelerate and even stop, until all pedestrians pass through the street. Meanwhile, small amount of cars and buses influence the operation of urban rail-bound transport, due to their temporary or scheduled stops in addition to the hindrance by pedestrians. Moving motorized road traffic may affect the following urban rail-bound transport. However, because both of them move carefully at relatively low speed along the shared space, the influences on urban rail-bound transport is very small too. At the shared space, after pedestrian pass through the cross or the stopped motorized road traffic restarts again, this hindrance is finished, though, other hindrances from road traffic may occur along the shared space at any location any time. It is different from the situation along the shared road. Once the urban rail-bound transport is blocked by road traffic, it will be hindered all through the shared road.

In conclusion, at the mixed traffic zone of level crossing, road traffic influences are reflected on the operation of traffic control signaling system at the level crossing. However, at the mixed traffic zone of shared road, various speeds of road traffic types have to be taken into account, especially, the lowest speed of all road traffic in front of the urban rail-bound transport. Randomness of pedestrians' behaviors, temporary stops by cars, or scheduled stops by buses have influences on urban rail-bound transport in the mixed traffic zone of shared space. The movements of urban mixed traffic that comply with the official traffic regulations at mixed traffic zones are default principle.

2.1.4   Special Conditions

For road traffic influences on urban rail-bound transport, various conditions of operation may affect the results. However, some unusual conditions will not be discussed in this dissertation, as described in the DFG-research project [Martin & Liu 2016a]. Such special conditions are summarized into six conditions with specific cases. As shown in Table 2-1, the special conditions include weather, technical hitches, non-compliance with traffic rules of road traffic, emergence cases, unpredictable accidents and special events, which are not discussed in this dissertation.

Special Conditions	Possible Cases/ Examples	Description
Weather	Sunny Day	The traffic loads and speeds of road traffic may be affected under various weather conditions.
	Rainy Day	
	Snowy Day	
Technical hitches	Infrastructure	It is possible that technical hitches occur on road traffic and urban rail-bound transport.
	Signal System	
	Rolling Stock	
Emergency cases	Ambulance	The vehicles of emergency always have the highest priority in case of emergency incidents.
	Fire Vehicles	
Non-compliance with traffic rules	Mishandling	Cases under this situation are totally out of control, which can to lead to dangerous and even accidents.
	Wrong Turnings	
Unpredictable accidents	Vehicle Accident	The accidents may disturb the operation of urban mixed traffic.
Special events	Major Match	Sporadic events are held under the control of relevant departments.
	Parade	
	Festival	

Table 2-1: Special or Unusual Conditions

Weather conditions can affect the trips of people, which can further impact traffic loads of road traffic and even the demand of urban rail-bound transport. In addition, weather conditions may also impact the speed of road traffic and urban rail-bound transport. The speeds of urban mixed traffic can consequently influence the urban rail-bound transport, especially on the shared road. Severe weather also leads to changes on travelers' trip purposes and the means of transport, resulting in the change of modal split on transportation. As described in the DFG-research project [Martin & Liu 2016a], various weather conditions, including the sunny day, rainy day and snowy day, may disturb the visibility of drivers of urban rail-bound transport. It's worthwhile to further investigate the relevant coefficients related to weather conditions, which can reflect the weather influences on operations of urban rail-bound transport in future research.

The technical hitches happen possibly on the rail-bound system in the aspect of infrastructure, signaling system and even rolling stocks, and also occur in vehicles and traffic signal systems of road traffic. For urban rail-bound transport, the technical hitches can be treated as a kind of perturbation along the rail tracks, which is not considered for evaluation of operational performance (recommended area of traffic flow introduced in Subchapter 2.2.3) of pure railway system with capacity research. The hitches of road traffic will further impact the operation of urban rail-bound transport, especially on the shared road. However, this condition is also an exceptional case, and will not be discussed in this dissertation.

Emergency situations exist inevitably in urban areas. The movements of all urban mixed traffic including urban rail-bound transport and road traffic should yield to the emergency vehicles, mainly fire trucks and ambulances. Such emergency situations are rare and are also excluded in this research.

Inevitably, some drivers or pedestrians will violate traffic regulations by taking wrong turns on the main road or passing through the urban roadways regardless under control of traffic control signals for example. Such behaviors are dangerous and are very likely to lead to accidents, disturbing the operation of urban mixed traffic. To avoid the collision with such private cars or pedestrians who violate traffic rules, urban rail-bound transport has to either reduce its speed or stop, ultimately disturbing the operation of urban rail-bound transport. The situation gets much worse if urban rail-bound transport has to stop on a sloping track because longer time will be needed to restart rail-bound transport in such situation.

Once in a while, some events are held, such as major matches or parades that involve with a very large group of population with a short notice. Usually, relevant departments adopt corresponding measures to deal with such special events.

Since all those situations are high random, these special conditions are not discussed in this dissertation. Nevertheless, it is meaningful and interesting to study the influences from these special conditions on urban rail-bound transport in future research.

2.2 Capacity Research for Urban Rail-bound Transport

[Pachl 2014] reported that capacity research is an important method to validate timetables of an existing or planned railway infrastructure and to design new infrastructure with adequate railway capacity, as well as to evaluate the operating performance of railway operation. Capacity research for urban rail-bound transport is carried out at mixed traffic zones, with the general basis summarized in this subchapter. The definition of relevant terms is introduced in Subchapter 2.2.1. There are two major methods for railway capacity research, which are respectively presented in Subchapter 2.2.2. Based on the methods of capacity research, the corresponding indicators can be derived as results for evaluation. The general structure of capacity research using simulation method is illustrated in Subchapter 2.2.3. This research is designed to evaluate the operating performance by analyzing capacity and operational quality of the urban rail-bound transport at various mixed traffic zones.

2.2.1 Basic Terms

Capacity research of urban rail-bound transport in this dissertation focuses on the operating performance of the investigated rail-bound system with mixed traffic zone(s). Some related basic terms are defined as following:

- **Operating Performance**: The relationship between the traffic flow of urban rail-bound transport and the quality of operation is used to determine the operating performance of the investigated rail-bound system with mixed traffic zone (confer [Pachl 2016]).

- **Traffic Flow (N in [trains/h])**: Traffic flow of urban rail-bound transport is the number of train runs as throughput in a time unit (one hour) during the investigated time period. During capacity research, no change is made on the given operating program in the investigated area (confer [Pachl 2016] & [DB Netz AG 2008]).

- **Operating Program:** The operating program in railway system is a comprehensive description of the operation process with the relevant requirements of throughput (confer [DB Netz AG 2008]). It includes the amounts of train runs, properties of trains, structure and sequence, as well as temporal allocation of the train runs.

- **Train Mixture:** train mixture is the structure of operating program, which can be treated as a rough operating program including the characteristics of various train groups[3] in the model and the allocation proportion of the number of trains in each train group.

- **Stepwise-varied Traffic Flows:** the stepwise-varied traffic flows of urban rail-bound transport are the corresponding traffic flows of stochastic timetables created by simulation method of capacity research (confer [Martin et al. 2011]). These created timetables are stepwise proportional to the traffic flow of an initial timetable[4] based on a given operating program. For example, if the traffic flow of an initial timetable based on a given operating program is 6 trains/h, a series of timetables with 50%, 100%, and 200% stepwise-varied traffic flows (3 trains/h, 6 trains/h and 12 trains/h) can be created. The number and proportion of stepwise-varied traffic flows can be determined based on the requirements, and each stepwise-varied traffic flow can create various timetables while keeping the structure of the given operating program.

- **Utility Factor of Capacity (η in [-]):** The utility factor of capacity is the ratio of the actual traffic flow divided by the (maximum) throughput capacity, which can be interpreted as the ratio of consumed capacity (confer [Chu 2014] and [DB Netz AG 2008]).

- **The Maximum (theoretical) Capacity ($MT\ LF$[5] in [trains/h]):** The maximum capacity is a theoretical value and is irrelevant to the requirement of operational quality, which allows unlimited congestion situation without keeping the structure of the given operating program. It corresponds to the maximum processable number of trains (including shunting movements) on the investigated infrastructure of the railway network through organization of train operation in

---

[3] Train group: Group of trains with same or similar characteristics, which operate on the same or similar routes.

[4] The initial time table is built in the simulation tool with detailed information for all investigated trains in the investigated time period. It usually refers to the actual operation, which can be used to determine the operating program in a selected time frame for creation of stochastic timetables with stepwise-varied traffic flows.

[5] $MT\ LF$ is the abbreviation of the term "Maximum (theoretical) Capacity" in German version of "Maximale (theoretische) Leistungsfähigkeit".

the process of scheduling in a given investigated time period. [DB Netz AG 2008]

- **Throughput Capacity ($DS\,LF$[6] in [trains/h]):** The throughput capacity proposed by [Chu 2014] is the average traffic flow of all possible maximum throughput per time unit (one hour) at static phase of operation process with various train sequences of a given rough operating program (train mixture). The maximum throughput (entry traffic flow = exit traffic flow) have to keep the structure of the defined rough operating program (train mixture) on a given investigated area. A further increase of traffic flow leads to a slow down on growth trend or a change of the given rough operating program (train mixture) of exit traffic flow.

- **Timetable Capacity ($FP\,LF$[7] in [trains/h]):** The timetable capacity was described in [Pachl 2016], which is the upper limit of capacity for timetable design. It is the maximum number of train path in the investigated area, which has to follow the given operational conditions, such as keeping the defined headway.

- **Investigated Area**: An investigated area is a defined as part of railway infrastructure, for which capacity research will be carried out. In this dissertation, the investigated area is a defined part of urban rail-bound network with one (an investigated mixed traffic zone) or more mixed traffic zones.

- **Investigated Time Period (T in [hour])**: An investigated time period is a defined time frame to carry out the simulation of capacity research. Generally, the investigated time period is about six hours (0:00:00 – 6:00:00) for simulation.

- **Evaluated Time Period (T_E in [hour])**: An evaluated time period is the time duration selected from the investigated time period. Generally, two hours in the

---

[6] $DS\,LF$ is the abbreviation of the term "Throughput Capacity" in German version of "Durchsatzbezogene Leistungsfähigkeit".

[7] $FP\,LF$ is the abbreviation of the term "Timetable Capacity" in German version of "Fahrplanleistungsfähigkeit".

middle of the investigated time period (2:00:00 – 4:00:00) are selected, which can ensure the evaluation of simulation results at a static state.

- **Recommended Area of Traffic Flow (OLB[8] in [trains/h])**: Recommended area of traffic flow is the range of traffic flow between the lower limit (left) with minimum value of relative sensitivity function of the waiting time and the higher limit (right) with maximum value of the so-called traffic energy function. It is an optimal interval of traffic flow to reach a customer-friendly quality of operation with the optimum utilization of a given infrastructure. (confer [Hertel 1992], [Schmidt 2009] & [Chu 2014]). In this dissertation, the recommended area of traffic is also an indicator of capacity research during the evaluation of urban rail-bound transport at mixed traffic zone.

2.2.2 Methods of Capacity Research

The dimension of infrastructure, including railway infrastructure and signal systems design, can be evaluated by the capacity research for railway system. Capacity research can also be utilized on operational planning under different demands. Optimization of the dimension of infrastructure and/ or the operating program is the way to improve the operating performance, which is one of the important objectives of capacity research. The waiting time of urban rail-bound transport used in describing the quality of operation and the corresponding capacity (traffic flow of urban rail-bound transport) is the essential indicator (results) for capacity research of performance evaluation. Capacity research for urban rail-bound transport at mixed traffic zones, determining the operating performance is also the most important evaluation objective to reflect the road traffic influences on the capacity and waiting time of the urban rail-bound transport in this dissertation.

Two primary methods of capacity research are used to determine the waiting time while evaluating the operating performance:

- Analytical method
- Simulation method

---

[8] OLB is the abbreviation of the term "Recommended Area of Traffic Flow" in German version of "Optimaler Leistungsbereich".

Analytical method is based on mathematic analysis. The capacity of the railway infrastructure and the waiting time are calculated using mathematical formula. Comparably, the simulation method is a method that models a real system or a process based on the considered characteristics. The evaluation results are statistical measures estimated by simulations for capacity research. Both methods can be used to determine the evaluation indicators for capacity research of railway systems, which will be introduced respectively in the following.

Analytical Method

The railway infrastructure and waiting time can be determined with analytical method through mathematical analysis with the characteristics of the given infrastructure and operating program ([Kontaxi & Ricci 2010] and [Pachl 2014]). In the study conducted by [Potthoff 1972], the queuing theory and probability theory were applied in analytical method. In order to determine the expected value of waiting time and the queue length with analytical method, the investigated area of railway system is modeled as an appropriate queuing system. Accordingly, the operating performance of the investigated area can be reflected through the relationship between waiting time (inverse proportion to the quality of operation) and capacity.

In such queuing system, rail track lines are modeled as single server while railway nodes (railway stations) are modeled as multi-servers using the analytical methods. A single server is utilized to model a subdivided section of rail tracks, which is relatively simple if only one train runs on one direction. Railway node is the connecting point of rail track lines in a railway network. The structures of railway nodes can be analyzed and divided into infrastructure elements, including set of tracks (Gleisgruppen) and node of conflicting sub routes[9] [Pachl 2016]. Based on the characteristics of the infrastructure elements, the set of tracks can be modeled as multi-servers in a queuing system. Comparably, single server is used to model a node of conflicting sub routes in a railway node, which is demarcated for exclusive route simultaneously. The single server can meet its characteristics. Accordingly, the expected value of waiting time is calculated for each node of conflicting sub routes during capacity research. Multi-

---

[9] Node of conflicting sub routes is the term "Teilfahrstraßenknoten" in German, which is referred to Railway Signalling Terms (English - German / Deutsch – Englisch), compiled by Hans Lindenberg.

resource queue is put into use to compile many nodes of conflicting sub routes in complex railway nodes (see [Omahen 1977], [Green 1984] and [Nießen 2008]).

With analytical method, the expected value of waiting time for a given operating program can be described mathematically. A proper probability distribution is used to describe the random variable of arrival times of trains (can be treated as customers in the queueing system). In addition, the service times of trains is treated as a random variable whose occupancy on the corresponding infrastructure elements can be calculated from a given probability distribution. With the mathematical model of infrastructure and operating program with analytical method, the expected value of waiting time and capacity can be further determined.

The compression method is used to determine the maximum capacity with the time-distance diagram in [International Union of Railways (UIC) 2004][10], which is also based on the analytical method. The blocking time stairways are compressed as closely as possible to estimate the consumed capacity of a railway line without changing the minimum line headway and the structure of trains in operating program.

The analytical method is strategically suitable for the evaluation of the railway system with a given operating program instead of a specific timetable. For relatively simple situations with homogeneous track arrangements and separated railway infrastructure sections, the analytical method is very useful in identifying the preliminary solutions and reference values for capacity research with mathematical formula and algebraic expressions, which needs much more efforts than the simulation method in terms of costs and time. (see [Pachl 2014] & [Rossetti 2009]).

Simulation Method

Simulation method is an experimental method, which is realized in a computerized simulation model of railway system to incarnate the real operation [Pachl 2014]. The simulation tools, such as RailSys[11], are developed to build a virtual laboratory with

---

[10] The compression method in [International Union of Railways (UIC) 2004] has limitation in usage. [Lindner 2011] demonstrated the problems and corresponding enrichment process for the restrictions of the sections division with same signal and infrastructure configuration as well as the same number of trains and train mixture for example.

[11] RailSys: Rail Management Consultants GmbH (RMCon), Hannover. [RMCON 2010]

the considered attributes of the investigated railway system. The simulation model can be used to study and evaluate the investigated railway system without many efforts. In addition, the virtual simulation model can virtualize a railway system with few limitations and is relatively more cost-efficient than an experimental model in reality [Siefer 2014].

The railway infrastructure, timetable and even perturbations can be set and incarnated in the proper model with the help of simulation tools. The accuracy of simulation results depends on the simulated model. The closer the simulation model is to reality, the better it can reflect the real operation.

There are two types of simulation methods, **synchronous simulation** and **asynchronous simulation**, based on the processing techniques with the help of suitable simulation tools. The operation process of trains in the model is simulated independently with the two types of simulation methods. For the synchronous simulation, all train runs in railway system are simulated simultaneously at each temporal point, based on the temporal arrival sequence to solve the conflict in the operation. Suitable simulation tools such as RailSys and OpenTrack[12] are used for synchronous simulation. In comparison, with the asynchronous simulation method, all train runs are based on their priority and sequence in succession while keeping the defined blocking time stairways. The conflicts are solved according to the priority of train runs. Train runs can be modeled in the operation process with the asynchronous simulation tool LUKS[13] for instance.

According to the simulation method, the waiting time function is an important intermediate result, which is based on the theory for capacity research of double-track railway lines [Hertel et al. 1987]. Further development by [Hertel 1992] proposed the waiting time function with two parameters and further derived the recommended area of traffic flow using analytical method. Based on the theory of [Hertel 1992], [Schmidt 2009] further improved the simulation method to determine the waiting time function and maximum capacity for complex and large-scale investigated area. The simulation

---

[12] OpenTrack: OpenTrack Railway Technology GmbH, Zürich. [Michl & Sojka 2014]

[13] LUKS: The software of railway simulation tool for research of an integrated system of railway operations, VIA Consulting & Development GmbH, Aachen. [Janecek & Weymann 2010a] and [Janecek et al. 2010b]

model with the investigated infrastructure and timetables based on operating program is established as the basis for capacity research. Through creation and simulations of stochastic timetables with stepwise-varied traffic flows semi-automatically with the help of simulation tool (like RailSys) and assistant software PULEIV[14], the relevant results can be derived.

Based on these research findings, [Martin & Chu 2012] and [Martin & Chu 2013] described the method to create stochastic influenced timetables more realistically in time slice. Accordingly, a series of timetables with stepwise-varied traffic flows were created for simulations to obtain a large sample of simulation results for the evaluation of operating performance in a cost-effective way. In this dissertation, consideration will be given to utilizing the synchronous simulation method for capacity research on urban rail-bound transport with road traffic influences.

2.2.3 Structure of Capacity Research

Capacity research is carried out for urban rail-bound system in mixed traffic zones using simulation method, which evaluates the operating performance through the quality of operation and capacity of the investigated area. The quality of operation is inversely correlated to the waiting time (i.e. operational delays) resulted from the simulated operation. The operating performance is generally characterized by the relationship of the quality of operation and the corresponding traffic flows (i.e. the number of urban rail-bound transports per hour) running in an investigated urban rail-bound infrastructure network.

---

[14] Programm zur Untersuchung des Leistungsverhaltens von Eisenbahninfrastrukturen, IEV, Stuttgart. It was an assistant software that was developed to create stochastic timetables with stepwise-varied traffic flows based on the operating program with consideration of the arrival delays distribution (perturbations of initial delays) [Martin et al. 2011].

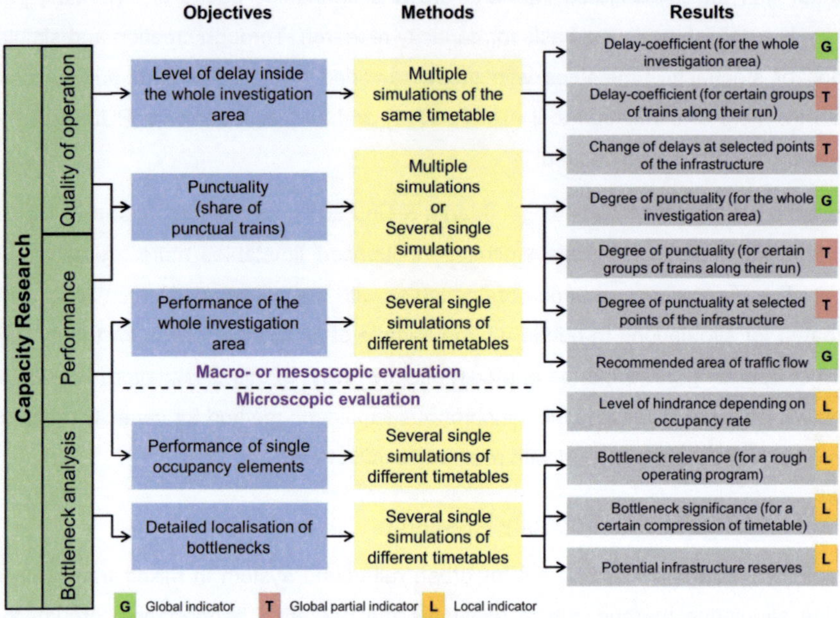

Figure 2-3: Structure of Capacity Research (Source: updated based on [Martin et al. 2012a])

The general structure of capacity research for railway system is shown in Figure 2-3. The capacity research on urban rail-bound transport in mixed traffic zones is an extension of railway system on the macroscopic or mesoscopic evaluation with road traffic influences. The performance of the whole investigated area is one of the objectives of capacity research designed to evaluate the quality of operation and the capacity of an investigated area with urban mixed traffic zone(s). Single simulations are mainly used to determine the result of recommended area of traffic flow to reflect the operating performance. In addition, the road traffic influences can be regarded as a kind of perturbation on urban rail-bound transport. Multiple simulations and several perturbed single simulations are used to determine the extra waiting time caused by road traffic.

There are two main intermediate results of capacity research to determine the operating performance: the throughput capacity for an investigated area with a given operating program with road traffic influences [Chu 2014], and the fitting curve of the adapted waiting time function (see Chapter 4) with data points (i.e. the waiting time concluded by simulations of each stochastic timetable). In Figure 2-4 the results of waiting time function and the throughput capacity are shown, which can further derive the recommended area of traffic flow to reflect the operating performance of the investigated area with a given operating program.

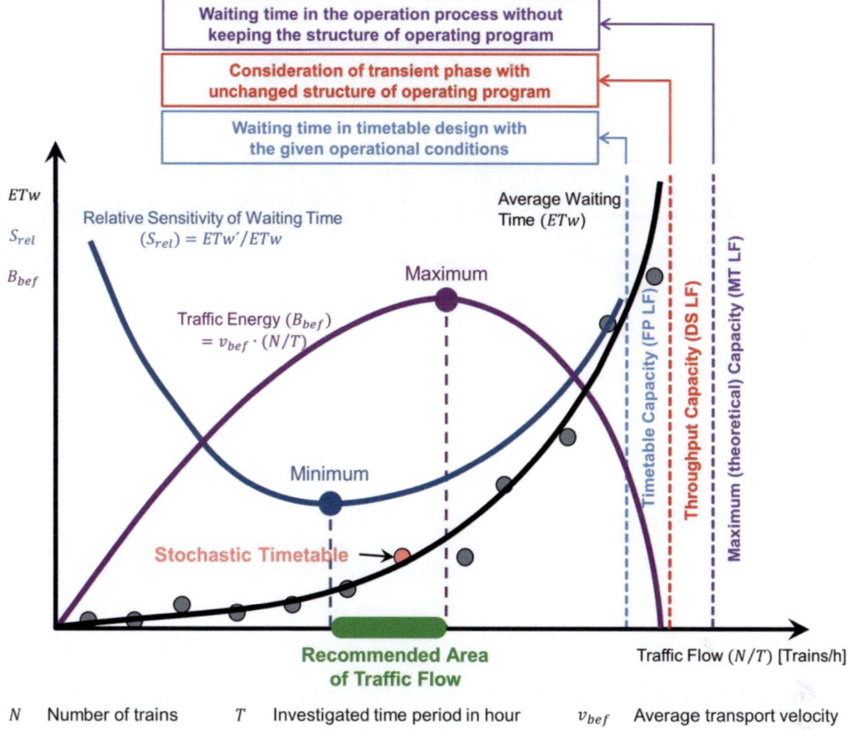

Figure 2-4: Evaluation of Performance for Capacity Research (Source: modified based on [Li 2015] & [Martin & Liu 2016a])

Throughput Capacity

The throughput capacity was proposed by [Chu 2014], from the view of analytical method, and it can be derived by the expected value of time interval between arrival time of two trains per time unit (one hour) $(1/ET_b)$. [Chu 2014] developed the simulation method based on [Schmidt 2009] (inflection point of entry traffic flow and exit traffic flow during the evaluated time period) to determine the throughput capacity (red dotted vertical line in Figure 2-4) with consideration of the transient phase effects with unchanged structure of the given operating program.

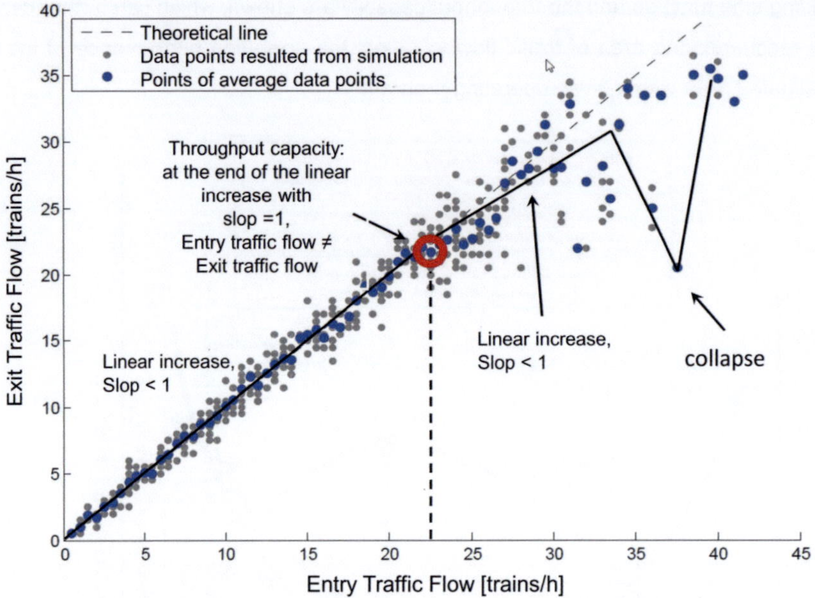

Figure 2-5: Relationship between Entry Traffic Flow and Exit Traffic Flow for Determination of Throughput Capacity (Source: modified based on [Chu 2014])

The systematic error between the determined inflection point of entry traffic flow and exit traffic flow and the real throughput capacity exists due to the consideration of transient phase as described in [Chu 2014] (Figure 2-5). With the consideration of the transient phase and the forerun time of simulation during the investigated time period, the throughput capacity can be calculated using:

$$M_r = \frac{M_e}{1 - \left(a_M \cdot \left(\frac{3600}{M_e}\right)^{c_M}\right)^{v^{b_M}}}$$

(2-1)

With:

M_r	The real throughput capacity ($DS\ LF$)
M_e	The determined inflection point of entry traffic flow and exit traffic flow
v	The forerun time of simulation in the investigated time period
a_M, b_M, c_M	The parameters of the model function of "Differenz" in [Chu 2014]

With the increase of the train traffic flow in the investigated area (x axis in Figure 2-4), the corresponding average waiting time increases to the throughput capacity, which is shown as the fitting curve of the waiting time function in black curve line in Figure 2-4.

Waiting Time Function

Some results from simulations such as transport time, entry and exit traffic flows and waiting time are recorded and further utilized to determine the throughput capacity and waiting time function. The recommended area of traffic flow for capacity research can be described based on the method in [Hertel 1992]. The waiting time function can express the relationship between the average waiting time and the traffic flows of trains, which represents the relationship between the quality of operation and the capacity of an infrastructure with a given operating program.

The black curve in Figure 2-4 is the fitting curve of data points (discrete waiting time) using the existing waiting time function. In order to reflect the randomness, a series of adequate samples of stochastic timetables with stepwise-varied traffic flows are created and simulated using Monte-Carlo-Method for simulation. Accordingly, data points from simulations of the stochastic timetables are summarized with average waiting time of each stepwise-varied traffic flow calculated to fit the curve based on the waiting time function, which will be discussed in Subchapter 4.2 of this dissertation.

Based on the derived waiting time function with the summarized data points (average waiting time of each stepwise-varied traffic flow), the recommended area of traffic flow can be further derived for the evaluation of operating performance.

Recommended Area of Traffic Flow

In order to find the region of the waiting time function that can reach an adequate quality of operation with an optimal traffic flow, the recommended area of traffic flow with two limit points of an "optimal region" is derived (the green frame region in Figure 2-4). The upper and lower limit points are the limits of the further derived curves based on the waiting time function (see [Hertel 1992], [Martin 2014] and [Pachl 2016]):

- **Lower limit** of the recommended area of traffic flow is defined as the minimum of waiting time with the relative sensitivity function (the dark blue curve in Figure 2-4), which can be interpreted as the minimal change (increase) of the average waiting time from the average waiting time of the increased traffic flow. The relative sensitivity $S_{rel}(\eta)$ of waiting time can be calculated from the quotient of the first derivative of waiting time function and the waiting time function itself.

- **Upper limit** of the recommended area of traffic flow is determined by the curve of the traffic energy function (the violet curve in Figure 2-4), which is the product of the number of trains (traffic flow) and the average speed per time unit. When there is no train in the system, the average speed is sure equal to zero. On the other hand, when the system is filled with trains (infinite close to throughput capacity), the waiting time goes up to infinite with the infinite reduced speed close to zero. Therefore, a maximum value of waiting time with the curve of traffic energy function is achieved with a certain number of trains at the corresponding speed.

It can represent the optimal relationship of capacity and quality of operation through the stochastic timetables in the investigated area within the recommended area of traffic flow. If the traffic flow is lower than the lower limit of recommended area of traffic flow, the value of waiting time is relatively low and increases moderately, which means the investigated area of railway infrastructure has lower utilization. On the contrary, if the traffic flow is higher than the upper limit of recommended area of traffic flow, the value of waiting time is high and will increase rapidly with a small growth of traffic flow. Under this situation, the high value of waiting time (delays) means the degradation of the quality of operation followed by a decrease of competition and accordingly a loss of customers.

2.3 Essential Tasks of Capacity Research for Urban Mixed Traffic

Capacity research is a comprehensive method as described in Subchapter 2.2. For the pure railway system, the operating performance is an important elevation objective of capacity research to reflect the relationship of capacity and quality of operation. The research on this topic is developed and improved step by step.

For urban mixed traffic zone, the structure of an investigated area may be complex and the extra influences caused by road traffic to consider are relatively random. Simulation method for capacity research is carried out in this dissertation depending on the existing research findings. The following difficult points in the research of urban rail-bound transport with road traffic influences have to be considered in this dissertation.

- How to reflect and describe the random road traffic influences at the investigated area with mixed traffic zone(s) in the simulation model with the help of simulation tool that is used to carry out capacity research?

- Various types of road traffic in various types of mixed traffic zones may result in different influences. How to calculate the road traffic influences (extra waiting time) on urban rail-bound transport and include such influences in the waiting time function and throughput capacity?

- In the mixed traffic zone, the urban rail-bound transport may cause waiting time not only through the hindrances between two trains but also the road traffic influences. Is the existing model function of waiting time function still suitable for the urban mixed traffic pattern?

- For simple and small-scale or relatively complex and large-scale investigated area, the (dis)advantages and the applicability of the developed methods have to be taken into consideration.

- If the required databases are not available or insufficient to carry out capacity research with simulation method for urban mixed traffic conditions, how to determine the results of capacity research for operating performance?

- For a whole investigated area with one or more mixed traffic zones, are all of them necessary to be studied? Is it possible that some of them may be insignificant with low impacts, which are unworthy a special study?

2.4 Objectives of the Dissertation

For the research on urban mixed traffic in this dissertation, further investigation in various urban mixed traffic zones based on the DFG-research project [Martin & Liu 2016a] are performed to evaluate the operating performance of urban rail-bound transport with road traffic influences (see Figure 2-6). The waiting time of urban rail-bound transport with the increase of traffic flows has to be determined. It is necessary to develop a general method to describe the waiting time of urban rail-bound transport with consideration of road traffic influences under different scenarios with various types of mixed traffic zones. In this dissertation, the challenges proposed in previous Subchapter will be analyzed preliminarily on the basis of the DFG-research project [Martin & Liu 2016a].

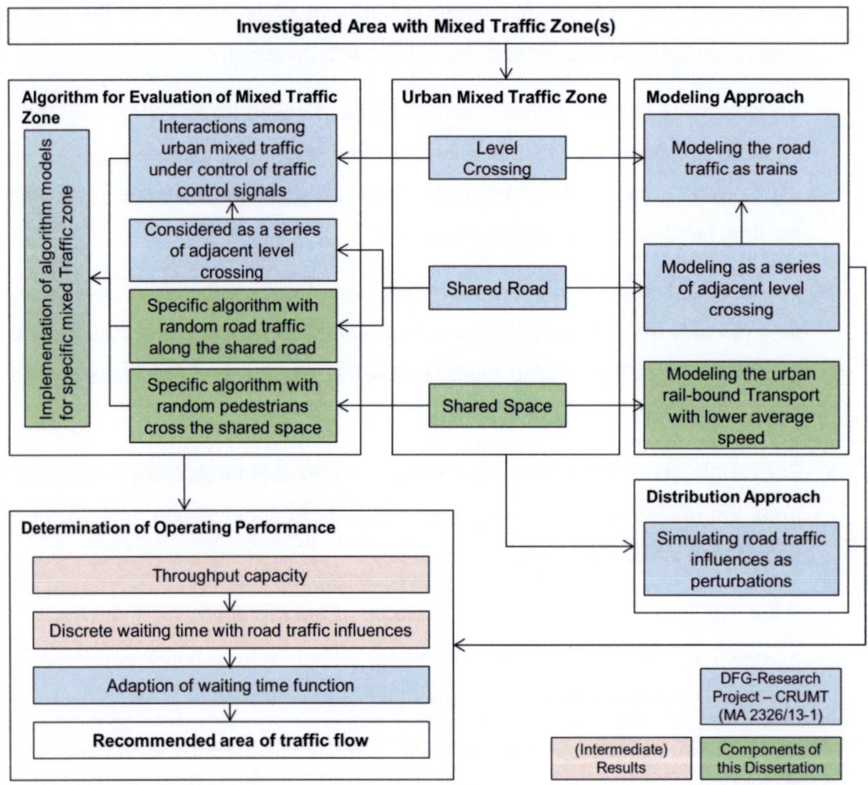

Figure 2-6: General Tasks

Firstly, based on the investigated mixed traffic zone, a simulation model with infrastructure and operating program is built for capacity research. Two approaches reported in the DFG-research project [Martin & Liu 2016a] are used to model the road traffic influences on urban rail-bound transport with the railway simulation tool Rail-Sys and assistant software PULEIV (see Chapter 3). Based on the basic concept, various types of road traffic in various types of mixed traffic zones can be reflected in different ways. The (dis)advantages and the applicability of them highly depend on the required efforts, databases, and the accuracy of results. The existing waiting time function used for urban mixed traffic deviates from the data points. Therefore, a new adapted model function of waiting time function is depicted on the basis of the DFG-research project [Martin & Liu 2016a] to achieve better fitting (see Chapter 4). To limit the efforts and required databases, an algorithm (see Chapter 5) will be further developed and expanded to derive the primary results of capacity research for various types of mixed traffic zones. In addition, it can also be used to assess the significance of mixed traffic zones from the impact point of view as the prerequisite of capacity research for a whole investigated area with one or more urban mixed traffic zones.

3 Approaches for Description of Road Traffic Influences on Urban Rail-bound Transport

3.1 Overview

Capacity research to evaluate the operating performance of pure railway system can be carried out based on simulation model built with railway infrastructure and timetable (operating program). For urban mixed traffic zone, except the multifarious processes of urban rail-bound transport in operation (similar to the railway operation), the road traffic in the investigated area cannot be overlooked. They are controlled by traffic control signals at mixed traffic zones of level crossing and shared road according to the relevant rules, and randomly move at mixed traffic zone of shared space, which can further influence the operation of urban rail-bound transport in mixed traffic zones as described in Chapter 2. In the DFG-research project [Martin & Liu 2016a] developed the following two approaches to describe the various road traffic influences systematically:

- Modeling Approach
- Distribution Approach

For modeling approach, the road traffic in the mixed traffic zone are modeled directly as trains in the rail-bound simulation model for modeling approach, while the perturbations of road traffic influences are added into the simulation model for distribution approach. The following subchapters describe the two approaches in terms of the description of road traffic influences on urban rail-bound transport in various urban mixed traffic zones.

3.2 Modeling Approach

3.2.1 Basic Concept

It is hard to express and quantify the road traffic influences on urban rail-bound transport with the increase of traffic flow in the simulation model of railway operation with the help of railway synchronous simulation tool, RailSys [RMCON 2010]. Moreover, although some simulation tools such as VISSIM[15] can be used to model the urban mixed traffic, the urban rail-bound transport in these simulation tools is modeled only for general purposes, which cannot fulfil the requirement of capacity research on urban rail-bound transport [Martin & Liu 2016a]. It is considered to model the road traffic as trains in the model of railway operation. In order to express precisely in the following context, two terms for this modeling approach are clearly expressed in the DFG-research project [Martin & Liu 2016a]:

- "Modeled road traffic" refers to the modeled trains in simulation tool representing the road traffic controlled by traffic control signals
- "Trains" refers to the trains modeled in simulation model that represent the urban rail-bound transports

The road traffic interfered with urban rail-bound transport in mixed traffic zones are modeled as urban rail-bound transports (trains) directly in the simulation model [Martin et al. 2012b]. Therefore, the influences caused by road traffic can also be directly reflected and described by the interactions (hindrances) between trains of "modeled road traffic" and "trains" of urban rail-bound transports in the simulation process [Martin & Liu 2016a]. Consequently, the influences of road traffic on urban rail-bound transports can be observed and investigated through the simulation with some restrictions.

3.2.2   Methodology of Modeling Approach

In this approach, the trains of "modeled road traffic" are simulated based on the operation of traffic control signals controlling the road traffic as described by [Martin & Liu 2016a]. Therefore, a big restriction with this approach is the applicability for the shared road and shared space. The shared road can be modeled as a series of fictitious connected level crossings along the road, which requires a lot of efforts, costs, and times. For shared space, the road traffic with random behaviors cannot be modeled as trains in the simulation. Each type of mixed traffic zone will be discussed in

---

[15] VISSIM: Verkehr In Städten – SIMulationsmodell, which can be used for simulation of multi-modal system and public transport in microscopic scale developed by PTV Planung Transport Verkehr AG in Karlsruhe, Germany. [Fellendorf & Vortisch 2010]

---

this subchapter respectively, which are further developed based on the DFG-research project [Martin & Liu 2016a].

Mixed traffic zone of level crossing

At a mixed traffic zone of level crossing, the movements of road traffic and even urban rail-bound transport are controlled by the traffic control signals at the level crossing. As described in Subchapters 2.1.3 and 2.1.2, the urban rail-bound transport competes with road traffic for the priority to pass through the level crossing. It is important to have both characteristics to build the simulation model with the integration of the trains of "modeled road traffic" [Martin & Liu 2016a]. Accordingly, the road traffic influences on urban rail-bound transport are similar to "trains" in simulation process.

- Topology structure of the investigated mixed traffic zone
- Representative estimation of "scheduled timetable" for trains of "modeled road traffic"

Regardless of rail-bound systems, the simulation model has to be built with the investigated infrastructure[16] and the timetable in the railway simulation tool RailSys. The interactions between "trains" of urban rail-bound transports and trains of "modeled road traffic" in the simulation can be described spatially by the topology structure in form of the layout of the investigated mixed traffic zone including rail-bound infrastructure and urban roadway (see Figure 3-1). For urban rail-bound transport, the corresponding rail-bound infrastructure for "trains" in the simulation tool can be built as common pure railway network for railway operation with all the relevant details such as rail tracks, switches, signaling systems and stations etc. as close to reality as necessary. On the other hand, for road traffic, a part of fictitious rail-bound infrastructure with the layout of the given urban roadways with the moving directions in the investigated mixed traffic zone for trains of "modeled road traffic" has to be built in the simulation tool [Martin & Liu 2016a].

---

[16] The infrastructure here indicates both of the rai-bound infrastructures for "trains" of urban rail-bound transports and trains of "modeled road traffic" running in reality on the urban roadways that are also modeled as rail-bound infrastructure in simulation model.

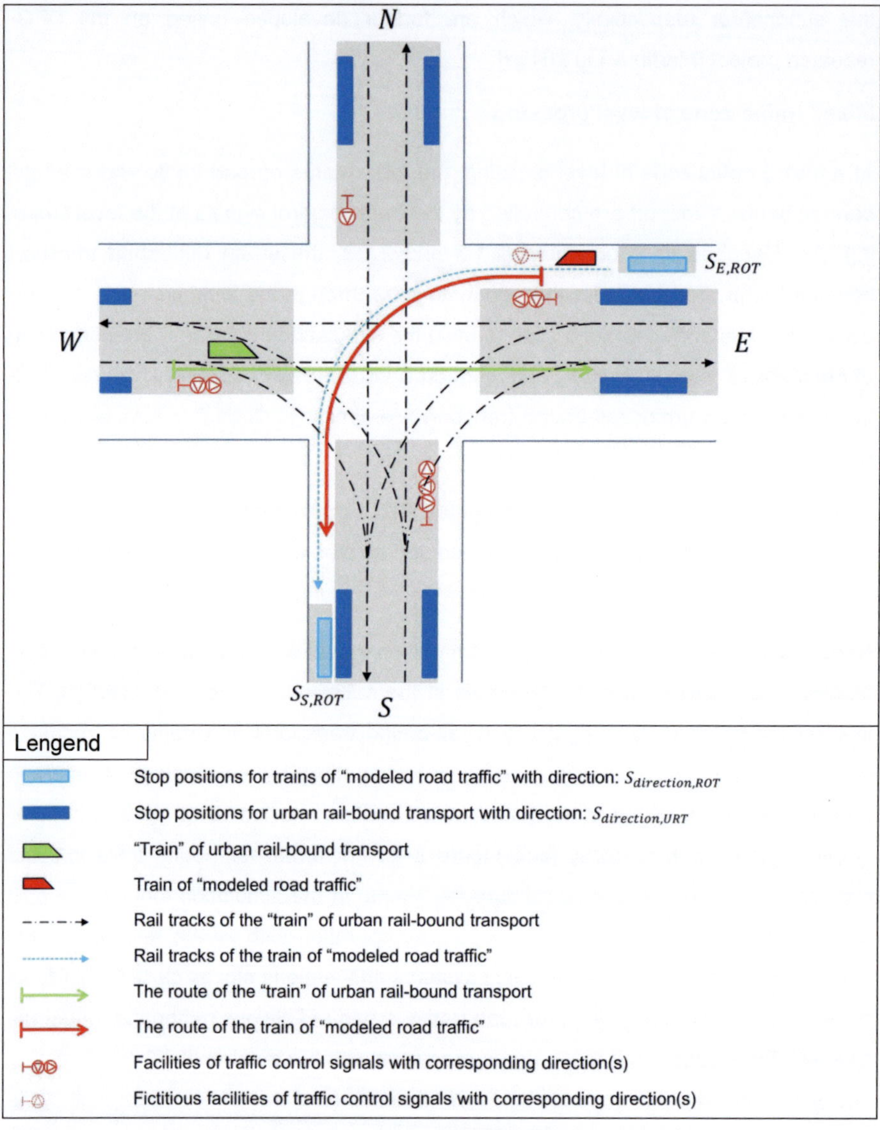

Figure 3-1: Representation of the Established Infrastructure of Mixed Traffic Zone of Level Crossing in Simulation Model (Source: modified based on [Martin & Liu 2016a])

The road traffic influences happen when the trains of "modeled road traffic" compete with the "trains" of urban rail-bound transports for the right to occupy the block of the level crossing. The traffic control signals control the movements of road traffic at the investigated level crossing. Therefore, the traffic light phases can represent the corresponding time durations of occupancy at the level crossing for road traffic to some extent. Integrated with the regulation of red light phase limitation, a rough frequency of road traffic influence is shown in the operation, which can be seen as the headway of the trains of "modeled road traffic" at the level crossing [Martin & Liu 2016a].

In addition, the other regulation of the minimum green light phase and the intervening time (from the end of a minimum green light phase to the begin of next green light phase in one traffic cycle) can be set as the corresponding blocking time of the trains of "modeled road traffic", which is the occupation time of road traffic under the controlled traffic control signal [Martin & Liu 2016a]. Therefore, the initial timetable for the trains of "modeled road traffic" can be built more accurately in the simulation model with the adjustment based on the collected data from the on-site observation.

Mixed traffic zone of shared road

Since the rail tracks of urban rail-bound transport share urban roadways, in the mixed traffic zone of shared road, they should also be utilized for trains of "modeled road traffic". In the DFG-research project [Martin & Liu 2016a], the basic concept was proposed for the modeling approach to evaluate shared road. The interactions between urban rail-bound transports and road traffic are not simple competitive relation similar to mixed traffic zone of level crossing.

The urban rail-bound transport depends on the movements of road traffic with different speeds (various road traffic modes) and the traffic loads along the investigated shared road. Moreover, the traffic control signals can only indirectly control the movements of urban mixed traffic at the end of the shared road, which are located at the next level crossing. Therefore, it is impossible to model the interactions between urban rail-bound transports and road traffic on shared road with the modeling approach totally same as at level crossing. Consideration is given to model the shared road along with a series of fictitiously connected level crossings.

Along the shared road, the urban rail-bound transport can be held back by road traffic at any point due to the low road traffic speed or the congestion of urban mixed traffic in front controlled by the traffic control signal at the next level crossing (see Figure 3-2 a) and b)).

If the urban rail-bound transport is hindered by road traffic with lower speed, it can only follow the road traffic with lower speed all along the shared road. On the other hand, if the urban rail-bound transport is held back by the congestion in front of it, the urban rail-bound transport can also run for the same time duration of green light phase with the road traffic in front when they proceed. The shared road can be modeled with the trains of "modeled road traffic" running at the fictitious level crossings with single rail track interfered with urban rail-bound transport.

The corresponding headway and blocking time are set based on the operation of the traffic control signal at the next level crossing (the traffic control signal at the end of the shared road in the direction of right in Figure 3-2 c)). The blocking time should be the non-green light phase (all time duration in traffic cycle rather than green light phase) of the traffic control signal at the next level crossing on the direction. The headway is similar to green light phase duration. It can be used to estimate the distance between two fictitious level crossings as running distance of urban rail-bound transport during the green light phase of the traffic control signal at the next adjacent level crossing.

In the simulation model, road traffic may hinder the urban rail-bound transport along the shared road randomly, which is reflected in the simulation process. It is possible that the urban rail-bound transport is not hindered at all or hindered several times along the shared road at any fictitious level crossing in the simulation. No hindrance along the shared road indicates that the urban rail-bound transport experiences no road traffic influences. If the urban rail-bound transport is hindered several times, these several hindrances should be counted from the end of the shared road backwards to the hindering point of the congestion of road traffic or the road traffic with lower speed.

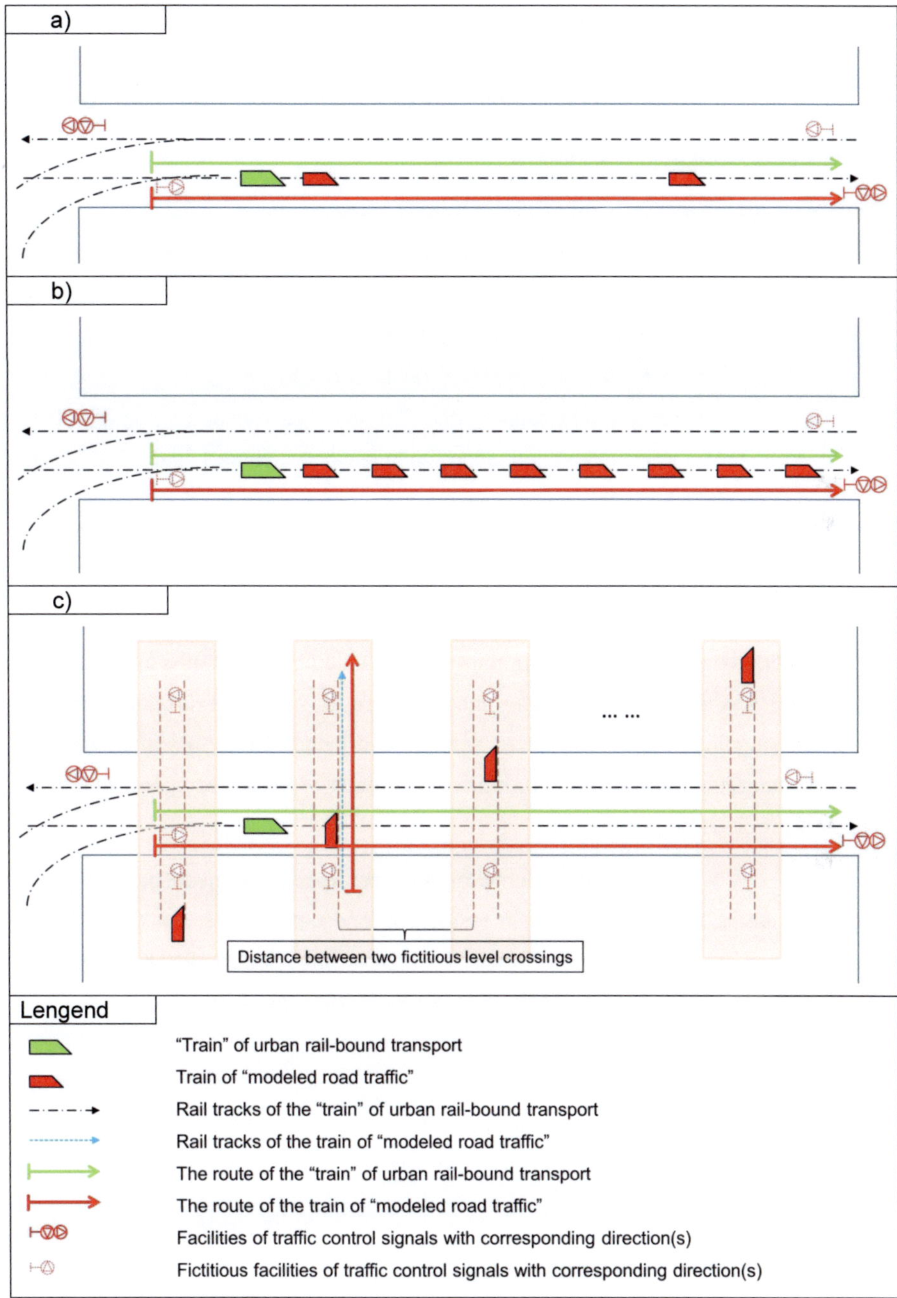

Figure 3-2: Established Infrastructure of Shared Road in Simulation Model

It is obvious that more time and higher cost are required to establish the fictitious level crossings and corresponding timetables for trains of "modeled road traffic" based on real operation for the evaluation of the shared road with this modeling approach.

Mixed traffic zone of shared space

Compared with the above investigated mixed traffic zones of level crossing and shared road, the situation in mixed traffic zone of shared space is totally random, without traffic control signals. With the modeling approach, it is hard to simulate the real road traffic by modeling pedestrians as "trains". Therefore, the road traffic influences are generally reflected on the urban rail-bound transport simply running at a lower average speed along the shared space. The average transport time per train t_b with or without road traffic influences along the investigated shared space can be determined respectively.

$$\Delta t_b = t_{b,with} - t_{b,without} \qquad\qquad (\text{3-1})$$

The difference of the two average transport time per train (Δt_b) in (3-1) is regarded as the waiting time of urban rail-bound transport caused by road traffic influences on the shared space. Therefore, even though there is only one urban rail-bound transport operating in the investigated area without hindrances between successive trains, some extra waiting time can occur as a result of the lower speed operation caused by road traffic influences.

In conclusion, the modeling approach can be utilized during the investigation of various mixed traffic zones under certain restrictions and deficiencies. For mixed traffic zones of level crossing and shared road, if a train of "modeled road traffic" occupies this investigated level crossing or fictitious level crossing on shared road, this (fictitious) level crossing will not available to other interfered "trains" of urban rail-bound transports. It shows that due to the operational hindrances between trains (both urban rail-bound transports and road traffic), the influences of road traffic on urban rail-bound transport can be reflected to a certain extent.

Further details of modeling approach were interpreted in the DFG-research project [Martin & Liu 2016a]. Firstly, when scheduling the timetables for both "trains" of urban rail-bound transports and trains of "modeled road traffic", it is important to reduce the possible conflicts between trains (both urban rail-bound transports and road traffic).

This can make the scheduled timetable much more plausible to be implemented. Secondly, setting priorities for "trains" of urban rail-bound transports and trains of "modeled road traffic" can eliminate some deficiencies in the scheduled timetable, which can simulate the road traffic influences much more reasonably. It is very helpful for the adjustment of train operation, especially when delays occur in simulation process. Furthermore, for the shared space, the setting for urban rail-bound transport with lower speed should be based on statistic data from the actual observations, so that the simulation results can be much more reliable.

In short, as the prerequisite for capacity research, the simulation model for urban mixed traffic using modeling approach is necessary and important. The urban rail-bound transport can be simulated with the modeled influences of road traffic in various investigated mixed traffic zones, which is much closer to reality. Therefore, the results of capacity research become more persuasive.

3.2.3   Capacity Research based on Modeling Approach

For capacity research, the throughput capacity and waiting time function are important intermediate results based on the simulation model, which can further derive the recommended area of traffic. In this subchapter, the capacity research is carried out based on the simulation model with modeling approach described in Subchapter 3.2.2, which was developed in the DFG-research project [Martin & Liu 2016a].

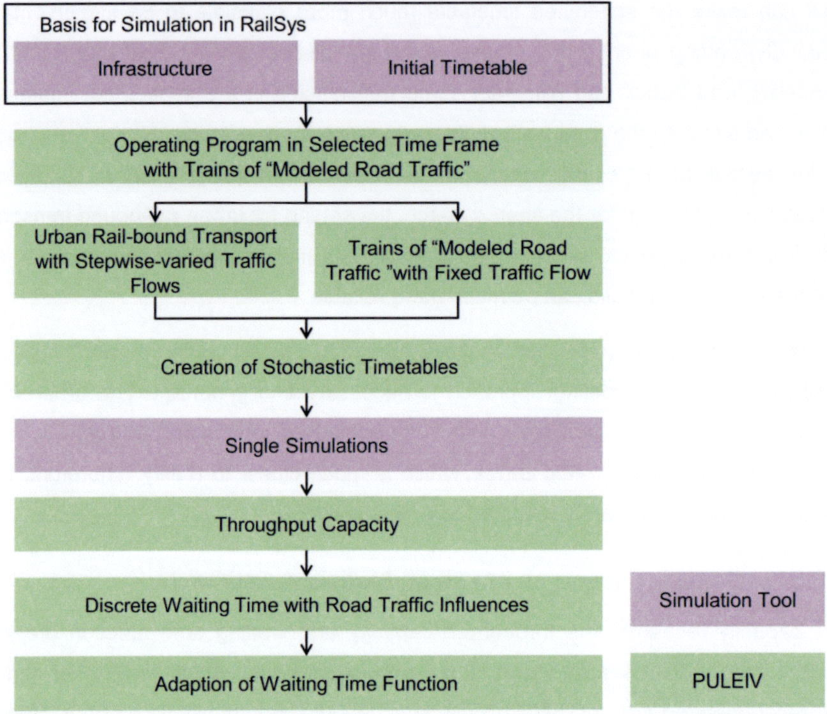

Figure 3-3: Workflow of Capacity Research based on Modeling Approach

Capacity research for urban rail-bound transport with road traffic influences is developed by the simulation method for common pure railway system. With modeling approach, the road traffic influences are already incarnated directly in the simulation process through the hindrances between successive trains ("trains" of urban rail-bound transports and trains of "modeled road traffic"). The whole workflow of capacity research based on modeling approach is illustrated in Figure 3-3.

Based on the initial timetable scheduled in simulation software RailSys according to the real operation, the operating program can be summarized with the help of assistant software PULEIV with both "trains" of urban rail-bound transports and trains of "modeled road traffic" in a selected time frame. The selected time frame is usually in the high traffic flow time period with higher traffic flow of urban rail-bound transport following the determined operating program.

Through the operating program in the selected time frame, a series of stochastic timetables with stepwise-varied traffic flows of urban rail-bound transport and fixed traffic flow of "modeled road traffic" can be created with the help of PULEIV. The reason that the traffic flow of "modeled road traffic" is fixed in these created timetables is because the road traffic influences in a selected time frame are only determined by the operation of traffic control signals for mixed traffic zones of level crossing and shared road. With the increase of traffic flow of urban rail-bound transport, the number of influenced urban rail-bound transport also increase. However, the traffic flow of "modeled road traffic" remains unchanged, and the increased road traffic influences on urban rail-bound transport are presented through the interactions between trains with the increase of the traffic flow of urban rail-bound transport in simulation process. Therefore, a series of stochastic timetables with stepwise-varied traffic flows of urban rail-bound transport and fixed traffic flow of "modeled road traffic" are created.

The single simulations can be executed in the simulation tool RailSys based on the built infrastructure and the created stochastic timetables. The results of simulations can be recorded and further summarized. The research is focused on the capacity and quality of operation of the urban rail-bound transport in investigated area with a given operating program. Therefore, the simulation result of waiting time is referred as the waiting time of "trains" of urban rail-bound transports with stepwise-varied traffic flows with consideration of road traffic influences. Consequently, the throughput capacity can be determined by the methods of [Chu 2014] and [Schmidt 2009].

Furthermore, the waiting time of "trains" of urban rail-bound transports with stepwise-varied traffic flows are recorded by simulation result. Therefore, the fitting curve of waiting time function can be derived with discrete data points that are the average waiting time of "trains" of urban rail-bound transports with each stepwise-varied traffic flow. Consideration of the existing waiting time function cannot be exactly appropriated for the urban rail-bound transport with road traffic influences. A new adapted model function of waiting time function is developed in this dissertation, which will be described in Chapter 4. According to the determined throughput capacity and the waiting time function, the recommended area of traffic flow can be further derived to show the operating performance.

Figure 3-4 shows the results of capacity research with road traffic influences based on the modeling approach compared with the results without road traffic influences. The blue curve is the fitting curve with the adapted waiting time function with road traffic influences (see Chapter 4). Comparably, the green curve with clear lower waiting time is the fitting curve with the waiting time function developed by [Chu 2014] without road traffic influences, which can be considered as a pure railway system.

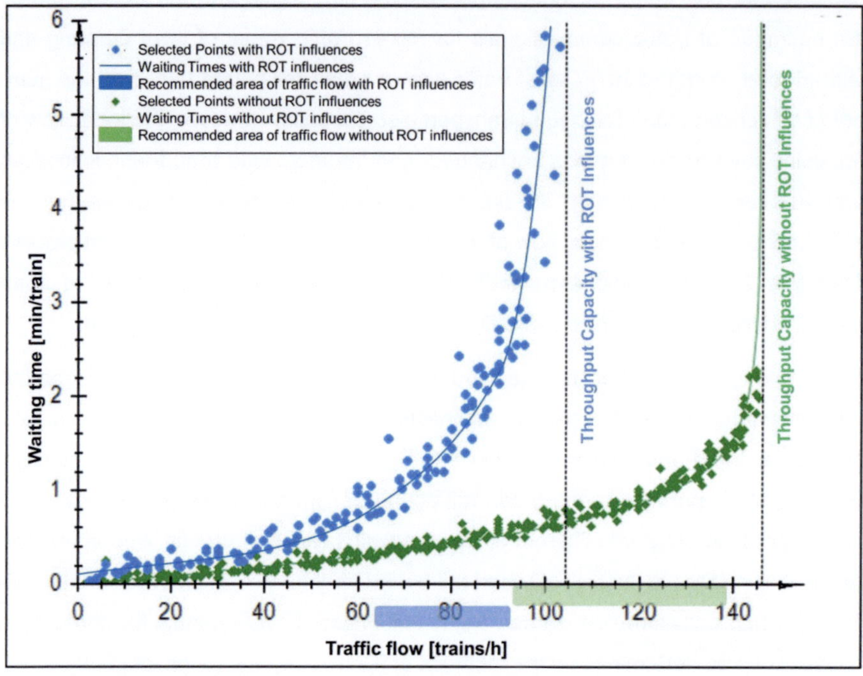

Figure 3-4: Waiting Time Function and Recommended Area of Traffic Flow based on Modeling Approach, compared with that without Road Traffic Influences[17] (Source: modified based on [Martin & Liu 2016a] & [Martin & Liu 2016b])

Additionally, the two throughput capacities are obviously different. Throughput capacity shows the number of urban rail-bound transport "maximally" running in an investigated area based on a given operating program when the waiting time towards infinite. The gap between the two throughput capacities indicates the road traffic influences. The road traffic restricts the possible capacity of urban rail-bound transport in the investigated area with urban mixed traffic zone(s). Similarly, the average waiting time increases with the decrease of operation quality, which is also caused by the influences of road traffic and leads to a lower capacity of urban rail-bound transport in

---

[17] It shows the capacity research results of an example of mixed traffic zone of level crossing introduced in Appendix I: Basic Information of the Investigated Example.

the investigated area. Consequently, the derived recommended area of traffic flow with road traffic drops obviously due to the road traffic influences.

3.3 Distribution Approach

Consideration of the application limit of modeling approach described in Subchapter 3.2, another approach is developed in this subchapter to overcome such insufficiency. The influences of road traffic can be regarded as the perturbations on urban rail-bound transport during the operation in the investigated mixed traffic zones, which can be added into the simulation model with suitable probability distribution. Therefore, this approach is called distribution approach as reported in the DFG-research project [Martin & Liu 2016a].

3.3.1 Basic Concept

For distribution approach, it is necessary to do the multiple simulations for perturbed urban rail-bound transport in the investigated area. The perturbed urban rail-bound transport is the urban rail-bound transport disturbed by stochastic influences of road traffic in the process of operation in the form of applicable mathematical distributions. As described in the DFG-research project [Martin & Liu 2016a], the road traffic influences as perturbations on urban rail-bound transport in the simulation model were introduced into the simulation tool RailSys. The impacts of the external influences of road traffic in mixed traffic zones can be analyzed in the form of delays (unscheduled waiting time for trains). The quality of operation can be evaluated and further optimized.

Perturbations are the disruptions to the scheduled timetable during operation, and perturbations are introduced into the urban rail-bound transport running with the scheduled timetable. Perturbation parameters[18] are utilized to describe perturbations in probability density functions. The corresponding possible probability distribution can be Erlang-k distribution, lognormal distribution, empirical distribution with actual collected data statistics, or the negative exponential distribution for multiple simulations in simulation tool RailSys etc.. When the urban rail-bound transport with the in-

---

[18] With the help of simulation tool RailSys, a perturbation is described by three perturbation parameters that will be introduced in Subchapter 3.3.2 in the form of probability density function in this dissertation.

serted perturbation of road traffic influences is simulated, the waiting time (delays) caused by road traffic influences in the operation process can be reflected in the simulation results directly. In this dissertation, negative exponential distribution and empirical distribution are used to describe perturbations on urban rail-bound transport with the simulation tool RailSys.

The inverse function of the cumulative distribution function based on the applicable probability distribution is utilized to define the delays caused by perturbations in simulation tool RailSys. Through a generated random number z[19], the probability of the occurrence of a delay of random number z can be used to derive the inverse function with z and the perturbation parameters [RMCON 2010]. In [RMCON 2010] gives the calculation of delays for a departure time extension as an example with the negative exponential distribution:

$$Del = \begin{cases} -del_m \cdot ln\left(1 - \dfrac{100z}{pro.}\right), & z < pro./100 \\ 0, & z \geq pro./100 \end{cases}$$

(3-2)

With:

Del	Delays caused by the perturbation of departure time extension
del_m	Perturbation parameter of average delays in minutes
$pro.$	Perturbation parameter of proportion values of perturbed urban rail-bound transport
z	The generated random number, with $z \in [0,1]$

Road traffic influences on urban rail-bound transport occur at mixed traffic zone and are caused by road traffic. Therefore, road traffic influences can be considered as one type of perturbation in the operation, which has to be determined for simulation. Besides, there are also other perturbations in various types on urban rail-bound transport that have nothing to do with the road traffic, but are related to the rail-bound system itself.

Various perturbations depend on the investigated time period. It is highly related to the operational situations regardless of road traffic influences or disruptions of rail-

---

[19] The random number z is in the interval of [0, 1], which can be generated even-distributed with a random number generator.

bound system itself. Using calibration algorithm (see Subchapter 3.3.2), the perturbation parameters of various perturbations in different types can be determined based on the data collected in real operation (from available statistic data or on-site measurement). Accordingly, the delays caused by various perturbations on urban rail-bound transports due to road traffic influences or disruptions of rail-bound system itself can be described in the simulation model (see Subchapter 3.3.3). Furthermore, the capacity research for urban rail-bound transport with road traffic influences can be carried out with the perturbations of road traffic influences (see Subchapter 3.3.4).

3.3.2   Perturbation Parameters

For urban rail-bound transport at the investigated area with one or more mixed traffic zones, there are different perturbation types to describe the disruptions on urban rail-bound transport caused by various reasons. The influences of road traffic can also be included into the simulation model as corresponding perturbation type, which leads to delays of urban rail-bound transport in operation. It is necessary to categorize perturbations that may result in various delays to the scheduled timetable on urban rail-bound transport for distribution approach. The perturbation types were introduced in the DFG-research project [Martin & Liu 2016a]:

- <u>Initial delay</u>: The delays already exist before urban rail-bound transports enter the investigated area. Because the perturbations occur outside the investigated area, the initial delays are introduced at the first stop of the whole investigated area in the simulation model.

- <u>Dwell time extension</u>: It is an unscheduled time extension of dwell time for boarding and alighting passengers at scheduled stop.

- <u>Running time extension</u>: It is also an unscheduled time extension of running time along the rail tracks that are delays of urban rail-bound transport caused by stochastic influences when train runs between two scheduled stops.

- <u>Departure time extension:</u> It is a time extension that also occur at scheduled stop when the boarding and alighting of passengers are completed, however, caused by technical failures of infrastructure or vehicles, or by driver's random behavior.

Correspondingly, there are three perturbation parameters: del_m, $pro.$ and del_{max}, for each perturbation type. It can be used to determine the delays with negative exponential distribution or empirical distribution, respectively [RMCON 2010]:

- del_m [minute]: Average value of delays.
- $pro.$ [%]: Proportion values of perturbed urban rail-bound transport (delayed urban rail-bound transport).
- del_{max} [minute]: Maximum delay for perturbed urban rail-bound transport

Based on the method developed in the DFG-research project [Martin & Liu 2016a], the road traffic influences are treated as a perturbation in the type of running time extension. In this dissertation, two running time extensions or departure time extensions can be categorized based on the causes. One is the described perturbation with the perturbation parameters P_{ROT}[20] caused by road traffic influences at urban mixed traffic zones, which occur along the rail tracks between two scheduled stops rather than at the scheduled stops. The other one with the perturbation parameters P_{URT}[21] is the disruption caused by technical disturbances, human reasons, or other reasons, which is the generalized running time extension (or departure time extension) of urban rail-bound transport regardless of road traffic.

Perturbation parameters (del_m and $pro.$) can be calibrated and determined with the calibration algorithm in [Cui et al. 2014]. Two indicators are the average value of delays (in minute) and the proportion of delayed trains (in %) respectively, collected by simulation results or actual operation. By comparing the given/ collected target values of indicators and the instantaneous values of indicators in iterative calibration process, the perturbation can be derived with the calibrated perturbation parameters. The target values of indicators are the reference values obtained from the actual statistic delays in the investigated area, which can be introduced into simulation model in the form of statistic distribution. Relatively, the instantaneous values of indicators

---

[20] P_{ROT}: Here indicates three perturbation parameters for running time extension (or departure time extension) caused by external influences of road traffic at mixed traffic zone.

[21] P_{URT} : Here indicates three perturbation parameters for running time extension (or departure time extension) caused by urban rail-bound system itself.

are the derived delays resulted from the outputs of multiple simulations based on the given inputs of the corresponding initiative values of the perturbation parameters.

The calibration process includes several rounds of multiple simulations, which is an iterative process. In multiple simulations, the corresponding perturbations of road traffic influences can be represented with the initiative value of the perturbation parameters set in each round. The values of these perturbation parameters are adjusted iteratively to minimize the difference between the output delays resulted from multiple simulations (instantaneous values of indicators) and the actual delays in operation (target values of indicators) to reach the convergence in the iterative calibration processes. The errors of each calibration round can be calculated and further minimized based on machine learning theory [Cui et al. 2014].

$$J(\gamma) = \frac{1}{2 \cdot 2 \cdot m_s} \sum_{i=1}^{m_s} \sum_{j=1}^{2} (v_{s,j}^{\gamma^{(t)}} - \overline{v_{s,j}})^2 \qquad\qquad (\,3\text{-}3\,)$$

With:

$J(\gamma)$	The error of the perturbation parameter $\gamma^{(t)}$
$\gamma^{(t)}$	Perturbation parameter in the t th round of calibration
$v_{s,j}^{\gamma^{(t)}}$	The instantaneous (normalized) value of j th indicator at stop s of the perturbation parameter $\gamma^{(t)}$
$\overline{v_{s,j}}$	The target (normalized) values of j th indicator at stop s
m_s	The number of stops

It is possible to derive the values of calibrated perturbation parameters P_{ROT} with the collected data of delays at scheduled stops for urban rail-bound transport in front and back of the investigated mixed traffic zone.

3.3.3 Methodology of Distribution Approach

Distribution approach is applied with sufficient statistic data collection. Through calibration process, the corresponding perturbation with the perturbation parameters P_{ROT} caused by road traffic influences can be determined for capacity research. The perturbation caused by road traffic can be modeled as additional running time extension (or departure time extension) with the perturbation parameters P_{ROT}. It is used for capacity research to determine the throughput capacity, fit the waiting time function with simulation results of waiting time, and further derive the recommended area of traffic flow with consideration of road traffic influences.

It is proposed in the DFG-research project [Martin & Liu 2016a], that the various perturbations (initial delay, dwell time extension, departure time extension and running time extensions) can be determined through calibration of the corresponding perturbation parameters based on the collected data with multiple simulations in the simulation model. The delays of urban rail-bound transport are caused by various perturbations simultaneously. In order to calibrate the perturbation parameters for running time extension (or departure time extension) with the perturbation parameter P_{ROT}, other various perturbations with theirs corresponding perturbation parameters have to be introduced into the simulation model as basis.

The perturbations of initial delay and dwell time extension on urban rail-bound transport can be introduced into the simulation model as inputs with the statistic data from real operation during the corresponding investigated time period[22]. The perturbation of initial delays of a given investigated area with mixed traffic zones can be obtained from actually collected data and set into the simulation model at the first scheduled stop. It is similar that based on the actually collected data at each specific scheduled stop, the perturbation of dwell time extension at each scheduled stop can be determined and introduced into the simulation model with proper statistic distribution.

The perturbation can also be obtained from the collected data. However, the statistical data of delays of urban rail-bound transport along the tracks is collected between two scheduled stops. It includes the perturbation of running time extension (or departure time extension) with the perturbation parameters P_{URT} that is the unscheduled time extension of urban rail-bound transport on rail tracks without consideration of external influences caused by road traffic, and the perturbation with the perturbation

---

[22] Generally, the perturbation varies during various time periods of operation. There are three time periods of operation in one day, which are considered: low traffic flow period, normal traffic flow period and high traffic flow period. For example, normally, during high traffic flow period the delays caused by various perturbations (initial delay, dwell time extension, running time extension and departure time extension considered in this dissertation) are relatively higher. On the contrary, it is the lowest delays caused by various perturbations during low traffic flow period.

parameters P_{ROT} that is the unscheduled time extension of urban rail-bound transport on rail tracks caused by road traffic. The waiting time of urban rail-bound transport caused by perturbation with P_{ROT} is the one needed for further capacity research without the waiting time caused by perturbation with the perturbation parameters P_{URT}.

It is hard to collect the data of delays if there is no road traffic influences in real operation in mixed traffic zones. And clearing away all road traffic is impossible. Therefore, an assumption that the operation of urban rail-bound transport at low traffic flow period (22:00 - 06:00), instead of no road traffic influence, is used in the distribution approach [Martin & Liu 2016a]. It assumes that the influences of road traffic on urban rail-bound transport at the low traffic flow period can be ignored for following calibration. Since the traffic load of road traffic is very low at this time period, it is assumed that no knock-on delay caused by road traffic either.

For different traffic flows of urban rail-bound transport during different time periods of operation, the differences of delays will be incarnated through the simulation process. If there is no perturbation, the (conflict free) timetable is simulated as the schedule without delays. The delays occur due to the hindrances between trains after the perturbations are introduced into the simulation model. If the traffic flow of urban rail-bound transport is low, the corresponding delays caused by introduced perturbations are relatively lower. On the other hand, an increase of traffic flows of urban rail-bound transport leads to the corresponding delays caused by introduced perturbations increasingly. With the same perturbations, the difference with various traffic flows can be reflected through the interactions of urban rail-bound transports in the simulation process.

Based on above assumption, there is one perturbation of running time extension (or departure time extension) with the perturbation parameters P_{URT} in addition to the initial delays and dwell time extension at low traffic flow period. It means that the delays of urban rail-bound transport caused by road traffic approach zero. Therefore, the perturbation with perturbation parameters P_{URT} can be derived from the calibration process with the collected data during low traffic flow period and the corresponding perturbations of initial delays and dwell time extension during low traffic flow period [Martin & Liu 2016a].

This perturbation with the perturbation parameters P_{URT} might be caused by technical problems, including failures of traffic control signaling system or trains and human errors such as drivers behaviors, etc.. It can occur similarly any time during the operation process. Therefore, the perturbation with the perturbation parameters P_{URT} remains constant for other investigated time periods when the running time extension (or departure time extension) with the perturbation parameters P_{ROT} also exists in the simulation model. The perturbation with perturbation parameters P_{URT} and the corresponding perturbations of initial delays and dwell time extension at other investigated time period are the basis to estimate the road traffic caused running time extension (or departure time extension) with the perturbation parameters P_{ROT} in the iterative calibration process.

In conclusion, there are mainly two steps developed in the DFG-research project [Martin & Liu 2016a] to determine the perturbation with the perturbation ters P_{ROT} caused by road traffic on urban rail-bound transport in simulation model with distribution approach for capacity research.

<u>Step 1</u>: Determining perturbation with the perturbation parameters P_{URT} without road traffic influences during low traffic flow period through iterative calibration process.

<u>Step 2</u>: Determining perturbation with the perturbation parameters P_{ROT} caused by road traffic influences during the investigated time period through iterative calibration process.

The whole procedure is based on the iterative calibration process with multiple simulations. Therefore, the investigated infrastructure and scheduled initial timetable for urban rail-bound transport set up in simulation tool are the basis of the simulation model for multiple simulations.

For the step 1, in order to determine the perturbation parameters P_{URT} for running time extension (or departure time extension), a calibration process for urban rail-bound transport without any influence of road traffic is carried out during low traffic flow period, which is shown in Figure 3-5. It is necessary to prepare the correlated statistic data for perturbation of initial delay at the first scheduled stop of each train run in the investigated mixed traffic zone and the perturbation of dwell time extension of each scheduled stop for calibration with multiple simulations based on the input data of the infrastructure and timetable.

In the calibration process, multiple simulations are executed with the given initial values of the perturbation parameters P_{URT} for the running time extension (or departure time extension) without road traffic influences for each iterative round. Therefrom, with the adjustment of the perturbation parameters P_{URT} for perturbation in simulation model, the multiple simulation results of output delays (instantaneous values of indicators) are compared with the actual delays (target values of indicators) iteratively in the calibration process, until the minimized error is reached. It is possible to determine the approximate values of the perturbation parameters P_{URT} for running time extension (or departure time extension) without road traffic influences.

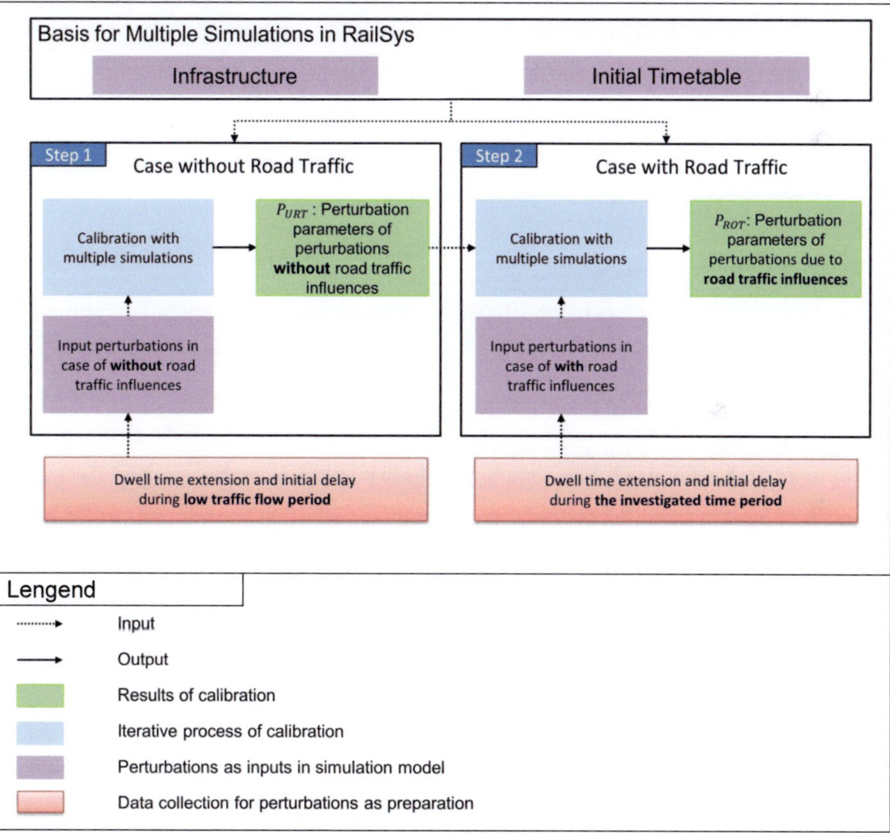

Figure 3-5: Procedure for the Determination of Perturbation Parameters P_{ROT} for the Running Time Extension or Departure Time Extension (Source: modified based on [Martin & Liu 2016a])

During the investigated time period, the delays caused by the perturbation of road traffic influences have to be taken into account. In step 2, the perturbation with the perturbation parameters P_{ROT} caused by road traffic can be derived based on the determined perturbation with the perturbation parameters P_{URT} from step 1.

Similarly, the calibration process can be carried out through multiple simulations with all the other perturbations, including the initial delay, dwell time extension of each scheduled stops in the investigated area collected from the available statistic data or on-site measured data during the investigated time period, and the derived perturbation parameters P_{URT} from step 1.

With the adjustment of initial values of the perturbation parameters P_{ROT} for running time extension (or departure time extension) caused by road traffic during the investigated time period iteratively in the iterative calibration process, the approximate values of the parameters P_{ROT} are further calibrated until the output delays of multiple simulations (instantaneous values of indicators) caused by both the rail-bound system and road traffic influences can match the delays collected in real operations (target values of indicators) with the minimized error.

3.3.4 Capacity Research with Distribution Approach

The perturbed single simulations [RMCON 2010] of stochastic timetables with step-wise-varied traffic flows are carried out for urban rail-bound transport with road traffic influences using distribution approach. The stochastic timetables are simulated with the perturbation with the parameters P_{ROT} caused by road traffic [Martin & Liu 2016a]. The extra waiting time of urban rail-bound transport exists already in the simulation results of waiting time. Accordingly, the operating performance with road traffic influences can be evaluated based on the simulation results.

The workflow of capacity research based on distribution approach is shown in Figure 3-6, which is based on the method for pure railway system without road traffic influences. Similarly, a series of stochastic timetables with stepwise-varied traffic flows have to be created with the assistant software PULEIV. As described in Subchapter 3.2.3, the investigated infrastructure and the initial scheduled timetable that determines the operating program are the basis for further simulation. These created stochastic timetables have to be set with the derived perturbation with perturbation parameters P_{ROT} caused by road traffic in the simulation tool. Each of these stochastic timetables with stepwise-varied traffic flows is perturbed with road traffic influences and can be simulated with the perturbation through perturbed single simulation.

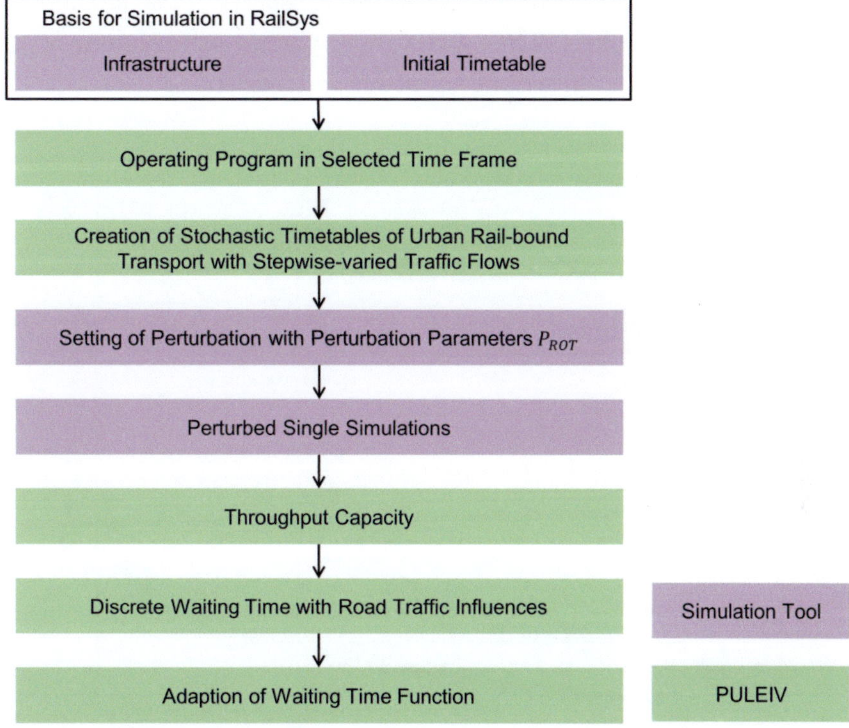

Figure 3-6: Workflow of Capacity Research based on Distribution Approach (Source: modified based on [Martin & Liu 2016a])

The simulation results of perturbed single simulations are summarized to determine the throughput capacity based on the method by [Chu 2014]. These results fit the curve of discrete waiting time (data points) with the developed waiting time function (see Chapter 4), and then derive the recommended area of traffic flow with road traffic influences to evaluate the operating performance.

The results of throughput capacity are similar to the results from modeling approach (see Figure 3-4). The results with two approaches are compared in Appendix II. A clear difference is observed between with and without road traffic influences. Figure 3-7 shows the results of capacity research with and without road traffic influences. The discrete average waiting time (data points in blue) based on the distribution approach includes the waiting time caused by road traffic influences, because all these simulated stochastic timetables are set with the perturbation of road traffic influences. Simulation results obviously include the extra waiting time due to influences of road traffic. The difference of waiting time between the data points in blue and in green shows the extra waiting time. Such extra waiting time occurs with the setting of perturbation with perturbation parameters P_{ROT} in simulation model.

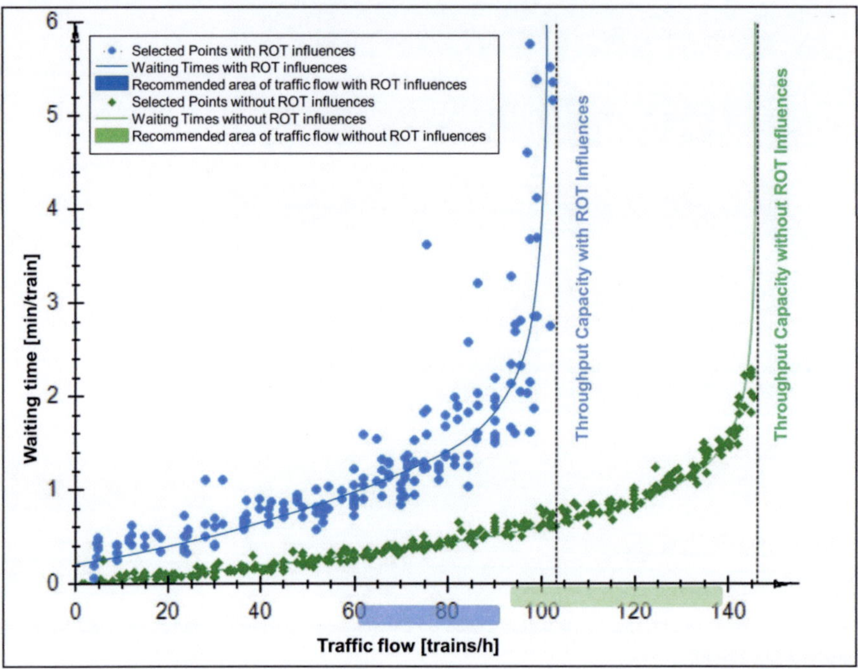

Figure 3-7: Waiting Time Function and Recommended Area of Traffic Flow based on Distribution Approach, compared with that without Road Traffic Influences[23] (Source: modified based on [Martin & Liu 2016a] & [Martin & Liu 2016b])

The blue fitting curve is the waiting time function (the adapted model function described in Chapter 4) with road traffic influences, while the green fitting curve is the waiting time function (the existing waiting time function used so far) for a pure railway system without road traffic influences. It is obviously that the road traffic influences restrict the capacity and lead to lower quality of operation. The difference between the two fitting curves can visually illustrate the influences of road traffic on urban rail-

---

[23] It shows the capacity research results of an example of mixed traffic zone of level crossing introduced in Appendix I: Basic Information of the Investigated Example, which is the same example to Figure 3-4.

bound transport. Consequently, the recommended area of traffic flow definitely lower due to the road traffic influences.

3.4 Conclusion

The two approaches described in this chapter are based on the DFG-research project [Martin & Liu 2016a]. The simulation model for urban rail-bound transport with road traffic influences can be built on two different ways. Modeling approach described in Subchapter 3.2 models the road traffic the same as the urban rail-bound transport as trains in the simulation model of rail-bound system. Adjusting the corresponding infrastructure and timetable to develop the integrated simulation model, the road traffic influences can be directly reflected on the hindrances between trains. For the mixed traffic zone of level crossing, the topology structure is relatively simple and the trains of "modeled road traffic" are under control of the traffic control signals. The interactions between urban rail-bound transports and road traffic are simple competition. However, for shared road, the road traffic affects the behavior of urban rail-bound transport, the interactions are not simple competition and the speeds of vehicle are different. It is hard to model such interdependent movements. As described in Subchapter 3.2.2, a lot of efforts have to be made on the study of shared road with the modeling approach. Furthermore, for the mixed traffic zone of shared space, it is also hard to model the flexible behaviors of pedestrians more precisely.

In summary, for a relatively simple and small-scale mixed traffic zone, modeling approach definitely has the advantage through modeling the road traffic as trains of "modeled road traffic" in simulation model with the help of simulation tool. No extra statistic data or efforts are required to establish infrastructure and timetable. However, for mixed traffic zones with large-scale or complex structures, especially where shared roads and shared spaces are involved, it costs a lot on data collection and the construction of infrastructure and timetable for simulation model with modeling approach. In addition, the results may be inaccurate due to the simplified simulation model.

Therefore, the other approach, distribution approach, described in Subchapter 3.3 is developed without restrictions on the types of mixed traffic zones. It is based on sufficient statistic data of delays of urban rail-bound transport, which is the big advantage of distribution approach. It is comprehensively usage for various types of mixed traffic zones and a whole investigated area with one or more mixed traffic zones. For a complex and confused or large-scale investigated area with one or more mixed traffic zones, it is unnecessary to recognize the specific conditions of each mixed traffic zone in the investigated area. However, the statistical data of delays of urban rail-bound transport caused by various perturbations have to be prepared, which is directly related to the cost, times, and accuracy of the results.

Furthermore, if there are several mixed traffic zones in an investigated area, some of them may have very small impact on urban rail-bound transport. If all of them have to be taken into account, it is unworthy and inefficient to make so many efforts. Therefore, an algorithm is developed based on the DFG-research project [Martin & Liu 2016a] to assess the significance of mixed traffic zones. It is the primary way to derive the results of capacity research when the two approaches are unavailable (see Chapter 5).

4 Modeling of an Adapted Waiting Time Function

4.1 Overview

The waiting time function plays a very important role in capacity research with analytical method or simulation method (see Subchapter 2.2.2) for evaluation of operating performance. It was proposed and developed for railway system step by step in the previous researches. In this dissertation, the coverage expands from the railway system to urban rail-bound system with mixed traffic zones. The extra waiting time caused by road traffic influences cannot be omitted in the waiting time function.

For urban mixed traffic pattern, the fitting curve of data points (discrete waiting time resulted from the simulated timetables with stepwise-varied traffic flows) with the existing waiting time function exhibits a deviation from the trend of them. The existing waiting time function is not appropriate for rail-bound system with urban mixed traffic zones. Accordingly, the accuracy of the further derived recommended area of traffic flow is somehow affected. Therefore, it is necessary to develop an improved model function of waiting time function for capacity research on urban rail-bound transport with road traffic influences. In this chapter, an improved waiting time function will be discussed based on the findings of the DFG-research project [Martin & Liu 2016a], which can be widely used to study the urban rail-bound transport with external influences.

4.2 Existing Waiting Time Function

With the development of capacity research, the waiting time function has been developed over the past a few years. The waiting time function was proposed by [Ludwig 1990] based on the mathematical principle of queuing theory. In [Schmidt 2009], the simulation method for capacity research was applied to the determination of the waiting time function with a logarithmic approximation approach based on the waiting time function developed by Ludwig [Ludwig 1990]. Further development on waiting time function for pure railway system without external influences using simulation method [Schmidt 2009] was studied by [Martin & Chu 2012], which considered the applicability on large-scale railway nodes or networks in reality. In addition, further consideration of the effects of transient phase in simulation process was discussed in

[Chu 2014], and an improved waiting time function was proposed based on the one developed by Ludwig [Ludwig 1990].

In this dissertation, the development on the waiting time function is based on the findings of [Chu 2014] and [Martin & Liu 2016a]. An adapted model function for waiting time function (see Subchapter 4.3) is developed for the operational situation of investigated area with mixed traffic zones, where the influences caused by road traffic on urban rail-bound transport exist.

4.2.1   Waiting Time Function by Ludwig

The first application of waiting time function for capacity research of railway system with double-track lines was reported by [Ludwig 1990] and [Hertel 1992]. Ludwig proposed the waiting time function with the assumption of queuing system $M/M/1$, which considers the stationary phase with two parameters [Ludwig 1990]:

$$ET_w = a \cdot \frac{\eta}{(1-\eta)^b} \qquad\qquad (\text{4-1})$$

With:

ET_w (statistical) Expected value of waiting time function

a, b Parameters of waiting time function

η Utility factor of capacity (traffic flow / maximum capacity)

The utility factor of capacity η is an important parameter in the waiting time function of any given infrastructure. It represents the amount of consumed capacity calculated by the quotient of the traffic flow in a specific timetable and the throughput capacity.

The queuing theory is the basic mathematical principle used to determine waiting time function. Based on the theory of Ludwig [Ludwig 1990], the railway infrastructure is modeled as a queuing system and trains are modeled as the customers (requests) in the queuing system of $M/M/1$. The arrival rate of trains specifies the average arrival frequency (rate) or the intensity per time interval in queuing system. It is indicated by the exponentially distributed random variables with parameter λ, which is the independent inter-arrival time of trains. The parameter μ is used to specify the service time that is also independent and exponentially distributed in the queuing system. ρ is the amount of consumed capacity calculated as the actual traffic flow divided by the maximum capacity in railway system, which is regarded as the parameter η in the waiting time function (4-1).

$$\rho = \frac{\lambda}{\mu} \qquad (\,4\text{-}2\,)$$

$$ET_w = \frac{\rho}{\mu(1-\rho)} \qquad (\,4\text{-}3\,)$$

With:

$\rho(\eta)$	Utility factor
λ	Average arrival frequency (rate)
μ	Average service rate of a single server

In the waiting time function (4-1), two parameters a and b were defined by [Ludwig 1990]. The coefficient of determination (R) can indicate the goodness of fit of a model function. It was analyzed and confirmed that the waiting time function (4-1) is suitable for simple railway structures such as partial track sections in [Schmidt 2009] and [Chu 2014].

The coefficient of determination (R^2) is defined as the difference between 1 and the ratio of the total sum of squares $SS(Total)$ and the residual sum of squares $SS(Res)$ with function (4-4) (see [Everitt 2002] & [Draper & Smith 1998]). The total sum of squares $SS(Total)$ is the sum, of the squares of the difference of the dependent variable and its overall mean. And the residual sum of squares $SS(Res)$ is the sum, of the squares of deviations between the data and an estimation model (residuals).

$$R^2 = 1 - \frac{SS(Res)}{SS(Total)} \qquad (\,4\text{-}4\,)$$

With:

R^2	The coefficient of determination
$SS(Res)$	The residual sum of squares
$SS(Total)$:	The total sum of squares

The coefficient of determination (R^2) is a useful statistical measure to check the regression fit to the real data points. It has a range of [0, 1] and is much closer to 1, which means that the regression line of waiting time function fits much better to the data points of investigated case.

Meanwhile, the suitable fit options of nonlinear robust least squares with regression method "bisquare" weights and Trust-Region-Algorithm[24] are used to increase the robustness and to better approximate the data points (the discrete waiting time) from simulation results by simulating the stochastic timetables with stepwise-varied traffic flows with waiting time function, which was discussed in the DFG-research project [Martin & Liu 2016a].

For large-scale railway nodes or railway networks with complex infrastructure sections, the capacity research has much higher practical significance. The waiting time function (4-1) exhibits some small deviations from the data points from simulation results.

4.2.2   Waiting Time Function by Schmidt and Chu

[Schmidt 2009] applied the simulation method on railway system to determine the maximum capacity and the waiting time function (4-1). As mentioned above, there are systematic deviations between data points of waiting time from simulation and the fitting curve with the waiting time function (4-1) [Chu 2014].

Consequently, a new model function of waiting time function was developed with consideration of transient phase in simulation process for railway system by [Chu 2014]. It can reach a relatively higher adaption for different railway networks and corresponding different operating programs. Accordingly, a much more reliable recommended area of traffic flow can be derived therefrom.

In order to study the effects of transient phase in the simulation process, [Chu 2014] put forward different model functions with different degrees of freedom[25] for waiting time function. By comparing coefficient of determination for different model functions

---

[24] Matlab 2014a [Moler 2008]

[25] Degrees of freedom: different model functions of waiting time function have different degrees of freedom. The waiting time function by Ludwig (4-1) with two variables has two degrees of freedom. Similarly the second-order polynomial function has three degrees of freedom and the third-order polynomial function has four degrees of freedom, which is even higher than the degree of freedom of the waiting time function (4-1).

of waiting time function, and further comparing adjusted coefficient of determination ($\bar{R}^2$) for different model functions of waiting time function with different degrees of freedom, it is able to estimate the adaption of approximation of fitting curve.

For the waiting time function with higher degrees of freedom, the adjusted coefficient of determination can be used to indicate the goodness of fit with different model functions of waiting time function, so that the suitable model function of waiting time function with higher explanatory power can be determined. The adjusted coefficient of determination ($\bar{R}^2$) is defined by the ratio of mean of squares $MS(Res)$ and $MS(Total)$ instead of sum of squares with function (4-5) [Draper & Smith 1998]. It is a rescaling of the coefficient of determination (R^2) by degrees of freedom.

$$\bar{R}^2 = 1 - \frac{MS(Res)}{MS(Total)} = 1 - \frac{(1 - R^2) \cdot (n - 1)}{(n_s - deg. - 1)} \qquad (4\text{-}5)$$

With:

$\bar{R}^2$	The adjusted coefficient of determination
n_s	The number of samples (data points)
$deg.$	The degrees of freedom of the waiting time function
$MS(Res)$	The residual mean of squares
$MS(Total)$	The total mean of squares

The model functions of waiting time function with Taylor expansion of second-order and third-order polynomial functions and exponential function were proposed in [Chu 2014]. However, the curve progressions of these proposed model functions of waiting time function cannot fit the data points from simulation (discrete waiting time) better than the waiting time function by Ludwig (4-1), which means there is no higher adaption level of approximation of fitting curve. Some improvements in [Chu 2014] and [Martin & Chu 2013] were reported on waiting time function based on the basic model function by Ludwig (4-1) with the consideration of effects of transient phase, which is much more suitable for large railway nodes and complex infrastructure sections of railway system in practice.

Two model functions of waiting time function were developed with three parameters by [Chu 2014] and [Martin & Chu 2013] respectively. The model function of waiting time function (4-6) was firstly developed by [Martin & Chu 2012] in which an extra parameter c_1 was inserted to increase the flexibility (degrees of freedom) of the waiting time function initially developed by Ludwig (4-1).

$$ET_w = a_1 \cdot \frac{\eta^{c_1}}{(1-\eta)^{b_1}} \qquad (\text{ 4-6 })$$

With:

ET_w (statistical) Expected value of waiting time function

a_1, b_1, c_1 Parameters of waiting time function

η Utility factor of capacity (traffic flow / throughput capacity)

The (adjusted) coefficient of determination of the new waiting time function (4-6) with three parameters and the basic waiting time function by Ludwig (4-1) with two parameters were compared, based on the fit option of nonlinear robust least squares of regression method "bisquare" weights and Trust-Region-Algorithm. The results showed that the waiting time function (4-6) has relatively higher (adjusted) coefficient of determination than the basic waiting time function (4-1) does, which means that the regression line of the waiting time function (4-6) with one extra parameter c_1 can fit better than the basic waiting time function (4-1).

Further development by [Chu 2014], a new model function of waiting time function (4-7) included the transient phase in simulation process for railway system, and has three parameters a_2, b_2, c_2.

$$ET_w = a_2 \cdot \frac{\eta}{(1-\eta)^{b_2}} \cdot (1-\eta^{c_2}) \qquad (\text{ 4-7 })$$

In this model function, an extra term, the third parameter $(1 - \eta^{c_2})$, is multiplied by the basic waiting time function by Ludwig (4-1), which is used to specify an approximation of transient phase in queuing system [Chu 2014] based on the theory of Ludwig. The number of trains in the transient phase and stationary phase is nearly the same as that with lower utility factor of capacity. However, the number reduces rapidly with the increase of utility factor of capacity. In [Ludwig 1990], the possible number of trains n_{req} is considered to be infinite in the stationary phase of the queuing system $M/M/1$. However, if the transient phase is taken into account, the possible number of trains in the queuing system $M/M/1$ is considered to be close to N_{req}.

$$EL_v = \sum_{n=0}^{\infty} n_{req} \cdot (1 - \rho) \cdot \rho^{n_{req}} \begin{cases} = \dfrac{\rho}{(1-\rho)}, & (n_{req} \to \infty) \\[2ex] \approx \dfrac{\rho \cdot (1 - \rho^{N_{req}})}{(1-\rho)}, & (n_{req} \to N_{req}) \end{cases} \qquad (\,4\text{-}8\,)$$

With:

EL_v — Average number of requests in queuing system

n_{req} — Possible number of requests in queuing system

$\rho(\eta)$ — Utility factor of capacity

The term $(1 - \rho^{N_{req}})$ reflects the relationship of average number of requests between transient phase and stationary phase. Consequently, [Chu 2014] proposed a model function of waiting time function (4-7) with parameter c_2 instead of N_{req} to express the effect of transient phase in simulation process.

The (adjusted) coefficient of determination is utilized to check the approbation of the proposed model function of waiting time function (4-7) with a multiply term $(1 - \eta^{c_2})$ with the same fit option of nonlinear robust least squares. [Chu 2014] summarized that the waiting time function (4-7) has a better coefficient of determination than the waiting time function (4-6), which can improve the goodness of fit with lower deviations from data points from simulation (discrete waiting time) than the waiting time function (4-1).

It is obvious that the waiting time function (4-7) can be widely used in railway system. However, if the extra influences are taken into account, extra waiting time likely increases for urban rail-bound transport in the investigated area with mixed traffic zones. Therefore, it is necessary to adjust the model function of the waiting time function for the special consideration of road traffic influences on urban rail-bound transport in this dissertation.

4.3 Terms of Adapted Waiting Time Function

An adapted waiting time function has to be developed when there are external influences on urban rail-bound transport in the investigated area. The waiting time of trains occurs even if there is no other hindered train. In a pure railway system, the waiting time is zero if there is no or only one train in the investigated area. However, for urban rail-bound transport in the investigated area with mixed traffic zones, even though there is only one train (urban rail-bound transport), the urban rail-bound transport may still be hindered by road traffic in mixed traffic zone, which lead to waiting time of urban rail-bound transport at mixed traffic zone theoretically.

Meanwhile, the waiting time caused by operational hindrances between successive trains is also taken into account except the external influences caused by road traffic, which have to be also specified in the adapted waiting time function. The adapted waiting time function was proposed based on the findings of the DFG-research project [Martin & Liu 2016a], which was based on the basic model function of waiting time function (4-7) developed by [Chu 2014] included the effects of transient phase in the simulation process.

4.3.1   Overview

In order to specify the urban rail-bound transport with road traffic influences, three adapted model functions of waiting time function were proposed in the DFG-research project [Martin & Liu 2016a]. All three model functions were based on the waiting time function (4-7) developed by [Chu 2014], and will be described in this subchapter. It is statistic result of the fitting curve of data points with the waiting time function (4-7). The starting point of zero indicates that the traffic flow of trains is in the range of (0, 1] in the pure railway system with a given operating program and that there is non-existent operational hindrance between two trains in the operation process.

For the urban rail-bound system with mixed traffic zone studied in this dissertation, the traffic flow of urban rail-bound transport in the range of (0, 1] also indicates the probability of only one train operating in the investigated area with a given operating program. It is possible that the waiting time does occur due to the operational hindrances between this urban rail-bound transport and road traffic in mixed traffic zone. Therefore, for urban rail-bound transport with road traffic influences, the starting point of the fitting curve for the adapted waiting time function should shift up slightly from zero, which shows the possible statistic waiting time of the traffic flow of urban rail-bound transport with the probability of one train. Hence, an additional parameter d is studied on the developed waiting time function to reflect the difference between results with and without road traffic influences.

It is valuable to analyze the new waiting time function with an additional parameter. The values of parameters in the adapted waiting time function can be approximated by the fitting curve of discrete data points with road traffic influences in simulations. The fit option of the nonlinear robust least squares with regression method "bisquare" weights and Trust-Region-Algorithm is also used to increase the robustness of the fitting curve of data points with the adapted waiting time function. It is necessary to assess and verify the fit of the new adapted waiting time function through the (adjusted) coefficient of determination.

4.3.2   Adapted waiting time function with a new constant term

The possible waiting time caused by road traffic influences in mixed traffic zone can be estimated when there is only one urban rail-bound transport in the investigated area. A constant term with parameter d_3 is considered to plus the waiting time function (4-7) developed by [Chu 2014] (the model function of the waiting time function (4-9) was firstly proposed by [Martin & Liu 2016a]). It is necessary to analyze the trend of the data points and the fitting curve with the adapted waiting time function with parameter d_3 (4-9) and to check the goodness of the fit through the (adjusted) coefficient of determination of the adapted waiting time function (4-9), whether it is higher than that of the existing one (4-7).

$$ET_w = a_3 \cdot \frac{\eta \cdot (1 - \eta^{c_3})}{(1 - \eta)^{b_3}} + d_3 \qquad\qquad (4\text{-}9)$$

With:

ET_w (statistical) Expected value of waiting time function

a_3, b_3, c_3, d_3 Parameters of waiting time function

η Utility factor of capacity (traffic flow/ throughput capacity)

The results show that the (adjusted) coefficient of determination of the adapted waiting time function (4-9) with a constant term of parameter d_3 is relatively higher than that of the existing waiting time function (4-7) as shown in Table 4-1. This indicates that for the investigated area with mixed traffic zones, the adapted waiting time function (4-9) with a constant term can fit the data points resulted from simulations slightly better than the existing one (4-7).

Waiting Time Function		Coefficient of Determination	Adjusted Coefficient of Determination
$ET_w = \dfrac{a_2 \cdot \eta \cdot (1 - \eta^{c_2})}{(1 - \eta)^{b_2}}$	(4-7)	0.9848	0.9847
$ET_w = \dfrac{a_3 \cdot \eta \cdot (1 - \eta^{c_3})}{(1 - \eta)^{b_3}} + d_3$	(4-9)	0.9877	0.9865

Table 4-1: Comparison of (Adjusted) Coefficient of Determination of the Adapted Waiting Time Function (4-9) and that of Existing Waiting Time Function (4-7)[26] (Source: [Martin & Liu 2016a])

The fit option is utilized to fit the curve, which can improve the robustness of the fitting curve. The parameters $(a_3, b_3, c_3$ and $d_3)$ in the adapted waiting time function (4-9) can be determined using the data points from the simulation. Figure 4-1 shows the results of fitting curves from the three model functions of waiting time function (4-1), (4-7) and (4-9). At the starting point, it is obvious that the waiting time function (4-9) with the added constant term of parameter d_3 can express the road traffic influences better than the other two model functions (4-1) and (4-7).

---

[26] The (adjusted) coefficient of determination shown here are the results of an example of mixed traffic zone of level crossing introduced in Appendix I: Basic Information of the Investigated Example.

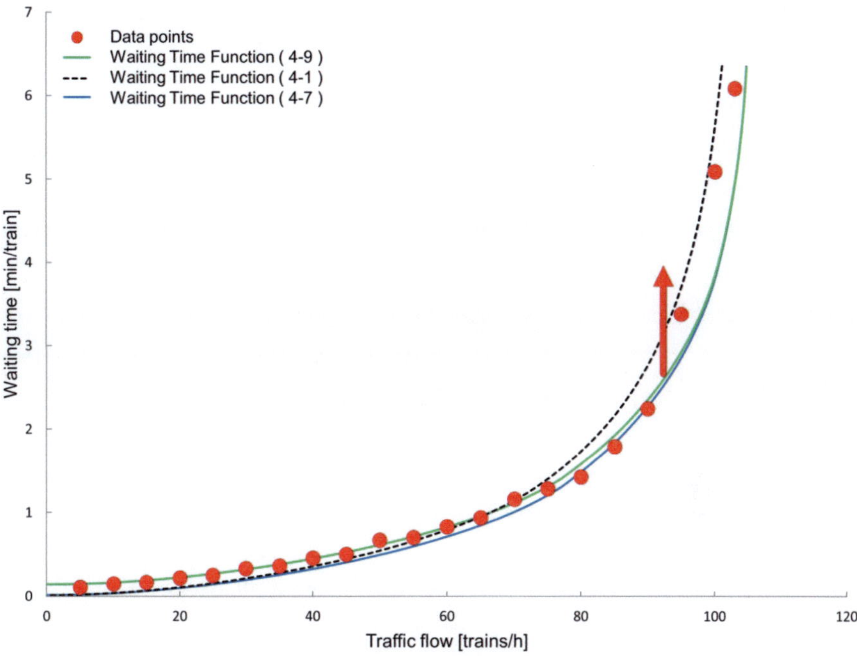

Figure 4-1: Fitting Curve with the Model Function of Waiting Time Function (4-9) with a Constant Term with Parameter d_3[27] (Source: modified based on [Martin & Liu 2016a])

However, it is also clear that there are some deviations between data points and fitting curve from the waiting time function (4-9). In order to improve the waiting time function, the comparison of the trend of data points and the fitting curve with the adapted waiting time function is analyzed. The difference between the fitting curve with the adapted waiting time function (4-9) and the data points shown in Figure 4-1 can be observed and the variation of data points (increase of average waiting time) with the increasing traffic flow of urban rail-bound transport is more rapidly than the fitting curve with the adapted waiting time function (4-9). It is especially obvious when the traffic flow of urban rail-bound transport is high.

---

[27] It shows the result of an example of mixed traffic zone of level crossing introduced in Appendix I: Basic Information of the Investigated Example.

4.3.3 Adapted waiting time function with new term related to traffic flow

A new model function of waiting time function (4-10) is developed with an additional term $d_4/(1-\eta)$. The fourth parameter d is based on the model function of waiting time function (4-9) with a constant term d_3. In order to show the increasing waiting time with the increase of the traffic flow of urban rail-bound transport, the new term $d_4/(1-\eta)$ was proposed in [Martin & Liu 2016a], and is proportional to the value of utility factor of capacity η that is related to the traffic flow of urban rail-bound transport. With increasing traffic flow of urban rail-bound transport, the value of the term $d_4/(1-\eta)$ increases much more rapidly than the constant term d_3 does.

$$ET_W = a_4 \cdot \frac{\eta \cdot (1 - \eta^{c_4})}{(1 - \eta)^{b_4}} + \frac{d_4}{(1 - \eta)} \qquad (4\text{-}10)$$

Similarly, the fit option of the nonlinear robust least squares and Trust-Region-Algorithm is also used to adjust the approximation of the waiting time function (4-10) with the data points. The green curve shown in Figure 4-2 is the fitting curve of data points with the adapted waiting time function (4-10). Comparably to the blue fitting curve of the waiting time function (4-9), the data points at starting point can be fit better with the waiting time function (4-10) with a new term $d_4/(1-\eta)$, a term that is not a constant parameter but related to the traffic flow of urban rail-bound transport.

As shown in Figure 4-2, the new waiting time function (4-10) fit the data points better than the waiting time function (4-7) and (4-9) with a constant term d_3. However, when the traffic flow falls in the range of middle to high, the fitting curve with the waiting time function (4-10) increases a little more quickly, which is a deviation between data points and the fitting curve.

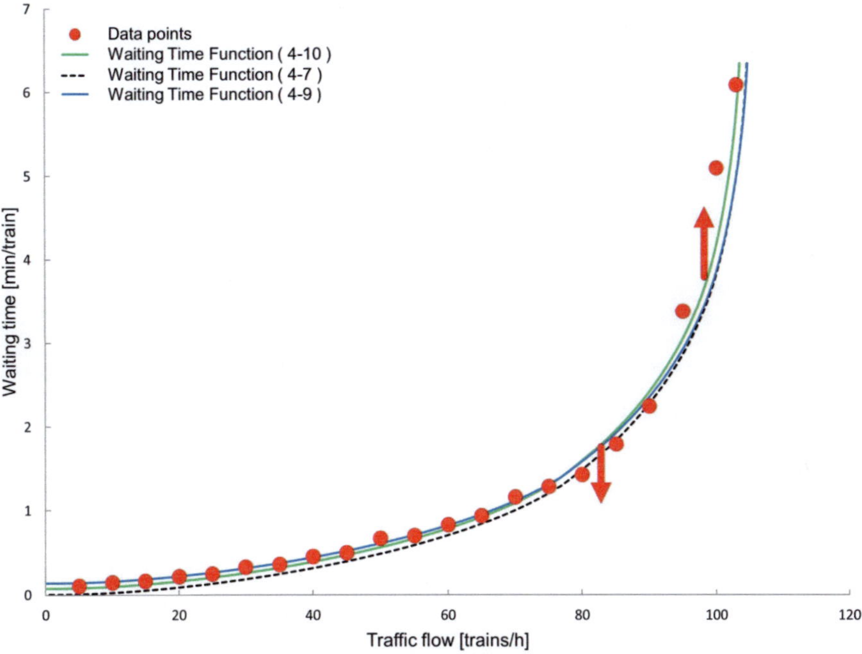

Figure 4-2: Fitting Curve with the Model Function of Waiting Time Function (4-10) with an additional Term $d_4/(1-\eta)$[28] (Source: modified based on [Martin & Liu 2016a])

It is clear that the of adjusted waiting time function (4-10) with the additional term $d_4/(1-\eta)$ can fit data points better than the existing one (4-7). Moreover, the adjusted waiting time function (4-10) also has a higher (adjusted) coefficient of determination than the existing one (4-7) as shown in Table 4-2 and the one with a constant term d_3 (4-10) (see Table 4-1), which means better fitting. Accordingly, the derived recommended area of traffic flow has a higher explanatory power.

---

[28] It shows the result of an example of mixed traffic zone of level crossing that introduced in Appendix I: Basic Information of the Investigated Example.

Waiting Time Function		Coefficient of Determination	Adjusted Coefficient of Determination
$ET_w = a_2 \cdot \dfrac{\eta}{(1-\eta)^{b_2}} \cdot (1 - \eta^{c_2})$	(4-7)	0.9848	0.9847
$ET_w = a_4 \cdot \dfrac{\eta \cdot (1 - \eta^{c_4})}{(1-\eta)^{b_4}} + \dfrac{d_4}{(1-\eta)}$	(4-10)	0.9901	0.9898

Table 4-2: Comparison of (Adjusted) Coefficient of Determination of the Waiting Time Function (4-10) and the Existing Waiting Time Function (4-7) [29] **(Source: [Martin & Liu 2016a])**

Based on the analysis of the different trend of data points and the fitting curve with the waiting time function (4-10) with the additional term $d_4/(1-\eta)$, for middle to high traffic flows of urban rail-bound transport, the fitting curve with the waiting time function (4-10) approximates the data points slightly faster. Therefore, consideration is given to lower the rate of increase of the waiting time with the waiting time function along with the increase of traffic flow. In the DFG-research project [Martin & Liu 2016a], a further adjusted model function of waiting time function (4-11) with an additional term $d_5/(1-\eta)^{2/3}$ was adapted.

$$ET_w = a_5 \cdot \frac{\eta \cdot (1 - \eta^{c_5})}{(1-\eta)^{b_5}} + \frac{d_5}{(1-\eta)^{2/3}} \qquad (4\text{-}11)$$

With:

ET_w (statistical) Expected value of waiting time function

$a_5, b_5, c_5 \ d_5$ Parameters of waiting time function

η Utility factor of capacity (traffic flow / throughput capacity)

---

[29] The (adjusted) coefficient of determination shown here are the results of an example of mixed traffic zone of level crossing introduced in Appendix I: Basic Information of the Investigated Example.

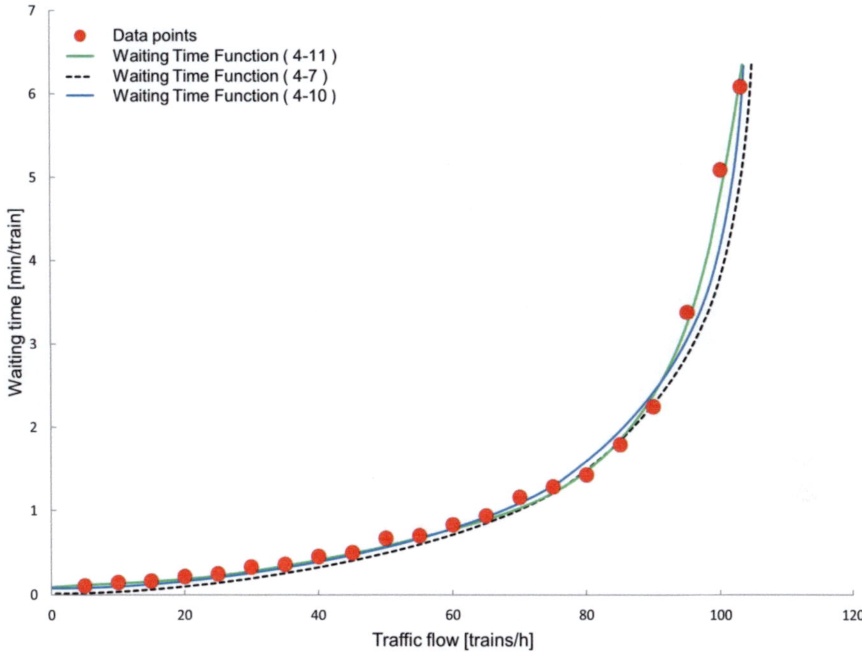

Figure 4-3: Fitting Curve with the Model Function of Waiting Time Function (4-11) with an additional Term $d_5/(1-\eta)^{2/3}$[30] (Source: modified based on [Martin & Liu 2016a])

As shown in Figure 4-3, there is almost no obvious deviation from the trend of the data points using the new adapted waiting time function (4-11). The new term $d_5/(1-\eta)^{2/3}$ adjusts the changing rate of the fitting curve from the starting point along with the increase of traffic flow of urban rail-bound transport. Compared with the fitting curve using the waiting time function (4-10), the curve from the new adapted waiting time function (4-11) with the added term $d_5/(1-\eta)^{2/3}$ can fit the data points better. Especially with middle to high traffic flows, the fitting curve of data points with the adapted waiting time function (4-11) can fit the data points very well.

---

[30] It shows the result of an example of mixed traffic zone of level crossing introduced in Appendix I: Basic Information of the Investigated Example.

Waiting Time Function		Coefficient of Determination	Adjusted Coefficient of Determination
$ET_w = \dfrac{a_4 \cdot \eta \cdot (1 - \eta^{c_4})}{(1 - \eta)^{b_4}} + \dfrac{d_4}{(1 - \eta)}$	(4-10)	0.9901	0.9898
$ET_w = \dfrac{a_5 \cdot \eta \cdot (1 - \eta^{c_5})}{(1 - \eta)^{b_5}} + \dfrac{d_5}{(1 - \eta)^{2/3}}$	(4-11)	0.9928	0.9927

Table 4-3: Comparison of (Adjusted) Coefficient of Determination of the Adapted Waiting Time Functions with Four Parameters (4-10) and (4-11) [31] (Source: [Martin & Liu 2016a])

Table 4-3 shows the (adjusted) coefficient of determination for two model functions of adapted waiting time function with the fourth parameter d. The results show that last adapted waiting time function (4-11) with the additional term $d_5/(1 - \eta)^{2/3}$ has relatively higher coefficient of determination and can fit the data points much better. Therefore, it is more plausible with higher explanatory power that the recommended area of traffic flow can be derived with the adaptive waiting time function (4-11). Compared with the results in Table 4-1, the two model functions of adapted waiting time function with the additional fourth parameter d (4-10) and (4-11) can reflect the changes of waiting time with the increase of traffic flow much better than the one (4-7) developed by [Chu 2014] with three parameters that has an insufficient (adjusted) coefficient of determination with road traffic influences. By contrast, the (adjusted) coefficient of determination for the one with an additional parameter d_3 (4-9) is relatively better.

4.4 Conclusion

The fitting curve of data points using the waiting time function has a high (adjusted) coefficient of determination based on the fit options of nonlinear robust least squares with regression method "bisquare" weights and Trust-Region-Algorithm, which means data points can be better fit and predicted with the waiting time function.

In this chapter, the adapted waiting time function (4-11) with an additional term $d_5/(1 - \eta)^{2/3}$ has a higher (adjusted) coefficient of determination than the waiting

---

[31] The (adjusted) coefficient of determination shown here are the results of an example of mixed traffic zone of level crossing introduced in Appendix I: Basic Information of the Investigated Example.

time function (4-7) developed by [Chu 2014] for rail-bound system with mixed traffic zones. The new model function can be used to describe the data points of average waiting time from simulation of the urban rail-bound transport with consideration of road traffic influences better. Accordingly, the derived recommended area of traffic flow for evaluating the operating performance is more plausible with higher explanatory power. Compared with other proposed model functions of waiting time function (4-9) and (4-10), the one with extra term $d_5/(1-\eta)^{2/3}$ can better represent the road traffic influences related to the traffic flow of urban rail-bound transport.

One more parameter d added in the adapted waiting time function makes it harder to fit the curve of adapted waiting time function in Matlab due to the higher degree of freedom. However, for the investigated area with mixed traffic zones, the influences caused by road traffic on urban rail-bound transport cannot be neglected during capacity research. It is necessary to use the adapted waiting time function with the consideration of external influences for deriving the recommended area of traffic flow to evaluate the operating performance.

5 Algorithm for Evaluation of Mixed Traffic Zone

As previously described, the simulation method for capacity research on pure railway system is mature. It is widely used to derive the waiting time function and throughput capacity through the waiting time form simulations of stochastic timetables with step-wise-varied traffic flows. Recommended area of traffic flow can be further determined to evaluate the operating performance. The waiting time in pure railway system without external influences caused by road traffic is a direct result of the hindered trains in operation process.

For an investigated area with mixed traffic zones, it can be studied using the approaches in previous chapters. However, due to the additional external influences of road traffic, it takes considerably higher efforts to derive the developed waiting time function (4-11) using the described approaches in Chapter 3 for capacity research. As described in Subchapter 3.4, both approaches can be applied with some restrictions. If the required information of road traffic is not fully available, it is hard to derive the new adapted waiting time function and throughput capacity with road traffic influences precisely [Martin & Liu 2016a]. Therefore, an algorithm needs to be developed to evaluate the operating performance of an investigated area with mixed traffic zone(s) in this chapter based on the findings of the DFG-research project [Martin & Liu 2016a].

5.1 Basic Concept

In order to derive the waiting time function (4-11) and throughput capacity with consideration of road traffic, this algorithm is developed to explore the way to acquire the external influences on urban rail-bound transport that is the extra waiting time caused by road traffic in operation of urban rail-bound transports in an investigated mixed traffic zone or even in a whole investigated area with one or more mixed traffic zones. Since it is difficult to acquire the detailed statistic data of road traffic influences in the operation in mixed traffic zone, models for various mixed traffic zones based on the developed algorithm with the basic road traffic information will be established to simulate the operation of urban mixed traffic in the investigated mixed traffic zone.

At mixed traffic zone of level crossing, the movement of road traffic is controlled by the traffic control signals as described in Subchapter 2.1. Therefore, the phase rotation of traffic control signals can reflect the operation of urban rail-bound transport and the movements of road traffic according to the pre-defined rules of traffic control signaling system. Similarly, the traffic control signal can also indirectly reflect the movements of road traffic and urban rail-bound transport on shared road. Moreover, the random location point where urban rail-bound transport is hindered by road traffic is described in the model. Comparably, in the shared space, the movements of pedestrians are totally random without traffic control signals. Therefore, in the model, the urban rail-bound transports in the shared space can be hindered at anytime and anywhere during the investigated time period.

In this chapter, the algorithm is developed in the model with an event-driven system, which represents the various cases responded to the triggered event that an urban rail-bound transport arrives at the investigated mixed traffic zone. Therefore, the extra waiting time of the urban rail-bound transport hindered by road traffic can be calculated. The algorithm is realized in the model with an automatic event-driven system, and enormous efforts can be save compared with the manual calculation [Martin & Liu 2016a]. In the basic model, the algorithm described in this dissertation is based on the simplified information that is summarized as the foundational rule of the operational road traffic. More detailed information of road traffic can be added into the model step by step with more acquired data as inputs for further development and practical application. The algorithm for basic model will be specified in Subchapter 5.2 with the relevant rules and input settings for various mixed traffic zones. Afterwards, the realization of the model in an event-driven system is described for various cases triggered by an event for various types of mixed traffic zones, which is discussed in Subchapter 5.3. Furthermore, some indicators for the determination of waiting time function and throughput capacity are interpreted in Subchapter 5.4. At last, Subchapter 5.5 will discuss how to determine the results of waiting time function and throughput capacity for various applicable goals and various investigated areas (a single mixed traffic zone or a whole investigated area).

5.2 Algorithm Description

5.2.1 Overview

The algorithm is developed to determine the road traffic influences with the increasing traffic flow of urban rail-bound transport in the investigated mixed traffic zone during the investigated time period. With the method of creation of stochastic timetables with stepwise-varied traffic flows [Schmidt 2009], the urban rail-bound transports are semi-randomly generated and included into the model based on the given operating program, which represents the urban rail-bound transports arriving at the mixed traffic zone with random initial delays [Martin & Liu 2016a].

This "semi-random" means that the generated urban rail-bound transports cannot be inserted total randomly during the investigated time period. However, considering the existing delays of urban rail-bound transport proceeding to enter the mixed traffic zone, the arriving time in operation cannot be exact as the one in the scheduled time-table. There is initial delay that needs to be taken into account. Therefore, the arriving time of urban rail-bound transports has to be generated randomly based on the basic scheduled timetable with stepwise-varied traffic flows. However, a determined time slice developed by [Chu 2014] is utilized to limit the randomness in order to keep the given operating program. The time slice (in second) is calculated using the investigated time period and the expected total number of urban rail-bound transports, which is shown as:

$$T_{sl} = \frac{T \cdot 60 \cdot 60}{\sum_{t=1}^{T} TF(t)} \tag{5-1}$$

With:

T_{sl} The time slice in second

T The investigated time period in hour

t The index of the investigated time period with $t \in [1, T]$

$TF(t)$ The expected traffic flow of the urban rail-bound transport per time unit (one hour) in the t th hour during the investigated time period

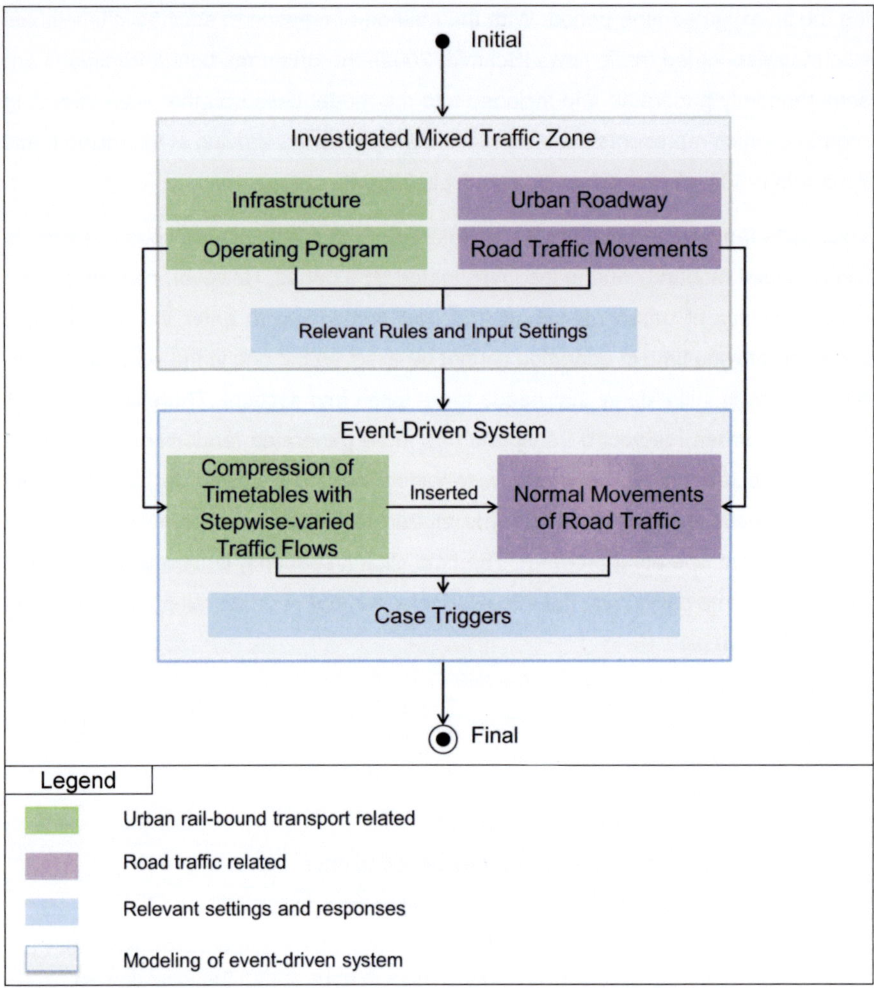

Figure 5-1: Scheme of the Algorithm Realized in the Model of Event-driven System (Source: modified based on [Martin & Liu 2016a])

Figure 5-1 illustrates the realization of the algorithm in the general process of modeling in an event-driven system. The basic information of the investigated mixed traffic zone is necessary and other prerequisites for modeling include the basic infrastructure, the given operating program (scheduled timetable) of urban rail-bound transport, and the movements of road traffic on the urban roadways. Meanwhile, the relevant rules have to be defined and set for the model as well.

Afterwards, the urban rail-bound transports with the corresponding stepwise-varied traffic flows are added into the normal movements of road traffic. An event will be triggered when an urban rail-bound transport arrives at the starting point of mixed traffic zone. Various cases are simulated to represent the interactions between the arriving urban rail-bound transport(s) and the road traffic in the event-driven system. Therefore, the influences of road traffic on urban rail-bound transport can be reflected in the model of event-driven system, which can be expressed as the hindrance time of urban rail-bound transport by road traffic in operation process.

5.2.2 Input Settings

The road traffic influences on urban rail-bound transport rise from the interactions between them. According to the conditions of various investigated mixed traffic zones, the interactions can result in different influences. Therefore, for various mixed traffic zones, some information as inputs for the modeling of the event-driven system has to be pre-defined respectively with this developed algorithm. In the DFG-research project [Martin & Liu 2016a], an important term **route** was defined for the developed algorithm:

"At mixed traffic zone, the **route** is defined as a section of infrastructure for urban rail-bound transport or an urban roadway for road traffic with direction, on which the urban mixed traffic (urban rail-bound transports and/ or road traffic) can move in the investigated mixed traffic zone from the entry point A to exit point B."

Routes can be subdivided into two sets in an investigated mixed traffic zone for urban rail-bound transport and road traffic respectively. However, the road traffic in shared space move without the control of traffic control signals. They are modeled as interfered points along the shared space instead of routes.

Route set of road traffic

$$R_{ROT} = \left\{ r_{ROT,1}, r_{ROT,2}, \cdots, r_{ROT,p}, \cdots, r_{ROT,m} \right\}$$ (5-2)

Route set of urban rail-bound transport

$$R_{URT} = \left\{ r_{URT,1}, r_{URT,2}, \cdots, r_{URT,q}, \cdots, r_{URT,n} \right\}$$ (5-3)

With:

$r_{ROT,p}$	A road traffic route in an investigated mixed traffic zone
$r_{URT,q}$	A urban rail-bound transport route in an investigated mixed traffic zone
m	The number of the road traffic routes with $|R_{ROT}| = m$
n	The number of the urban rail-bound transport routes with $|R_{URT}| = n$
p	The index of the road traffic routes, $p \in [1, m]$
q	The index of the urban rail-bound transport routes, $q \in [1, n]$

For an investigated mixed traffic zone, the routes have to be definite as inputs for modeling, allowing the corresponding urban mixed traffic to operate. The routes for urban rail-bound transport and road traffic in mixed traffic zones (except shared space) are illustrated with examples as shown in Figure 5-2. A summary of three types of mixed traffic zones is shown in Figure 5-2 a) with blue, pink and green frame respectively.

Further description in microscopic scale at mixed traffic zone of level crossing is shown in Figure 5-2 b). The location points of facilities of traffic control signals at the level crossing are set as the entry points (start points) of the possible routes in the direction entering the level crossing. Accordingly, the end point of a route is regarded as the point of the next adjacent traffic control signal in the direction, or can be set as an virtual traffic control signal at the end of the block based on the required signaling block system. The urban rail-bound transport route with the direction ($W \rightarrow E$) is defined as the section of infrastructure at the level crossing from the start point s_1 to the end point s_4 (virtual traffic control signal) that can be denoted as $r_{URT}: s_1 \rightarrow s_4$. Similarly, for road traffic route denoted as $r_{ROT}: s_5 \rightarrow s_2$ is a section of urban roadway with direction ($E \rightarrow S$) from the start point s_5 to the end point s_2 (virtual traffic control signal).

The mixed traffic zone of shared road is shown in Figure 5-2 c). Similar to that of level crossing, the location point of (virtual) traffic control signal is considered as the start point of a route. The end point of the route on shared road is the location point at the facility of traffic control signal in the direction at the following adjacent level crossing, which can be the start point of another route. In this example, the road traffic route on the shared road with direction ($E \rightarrow W_1$) is denoted as $r_{ROT}: s_4 \rightarrow s_9$. It starts from the end point s_4 of the route $r_{ROT}: s_7 \rightarrow s_4$ at the previous adjacent mixed traffic zone of level crossing to the end point s_9 that may be the start point of another route at the next level crossing. Due to the attribute of "sharing" on shared road, the road traffic share the urban roadway with rail tracks for urban rail-bound transport along the shared road. The urban rail-bound transport route is denoted as r_{URT}: $s_4 \rightarrow s_9$ also with start point s_4, end point s_9, and direction $(E \rightarrow W')$ in this example. Figure 5-2 d) illustrates an example of mixed traffic zone of shared space with the urban rail-bound transport denoted as r_{URT}: $s_{11} \rightarrow s_1$. The urban rail-bound transport route can also be defined with the start point of s_{11} and the end point of s_1 in the direction $(E_2 \rightarrow W)$. In shared space, road traffic are the pedestrian walk with random behaviors and the motorized road traffic (cars and buses) with low traffic load and speed. In this algorithm, pedestrians cannot be defined to move on fixed routes. Urban rail-bound transport on the route r_{URT}: $s_{11} \rightarrow s_1$ can be interfered at any time point and any location by random movements of pedestrians, temporary stops of cars, or scheduled stops of buses.

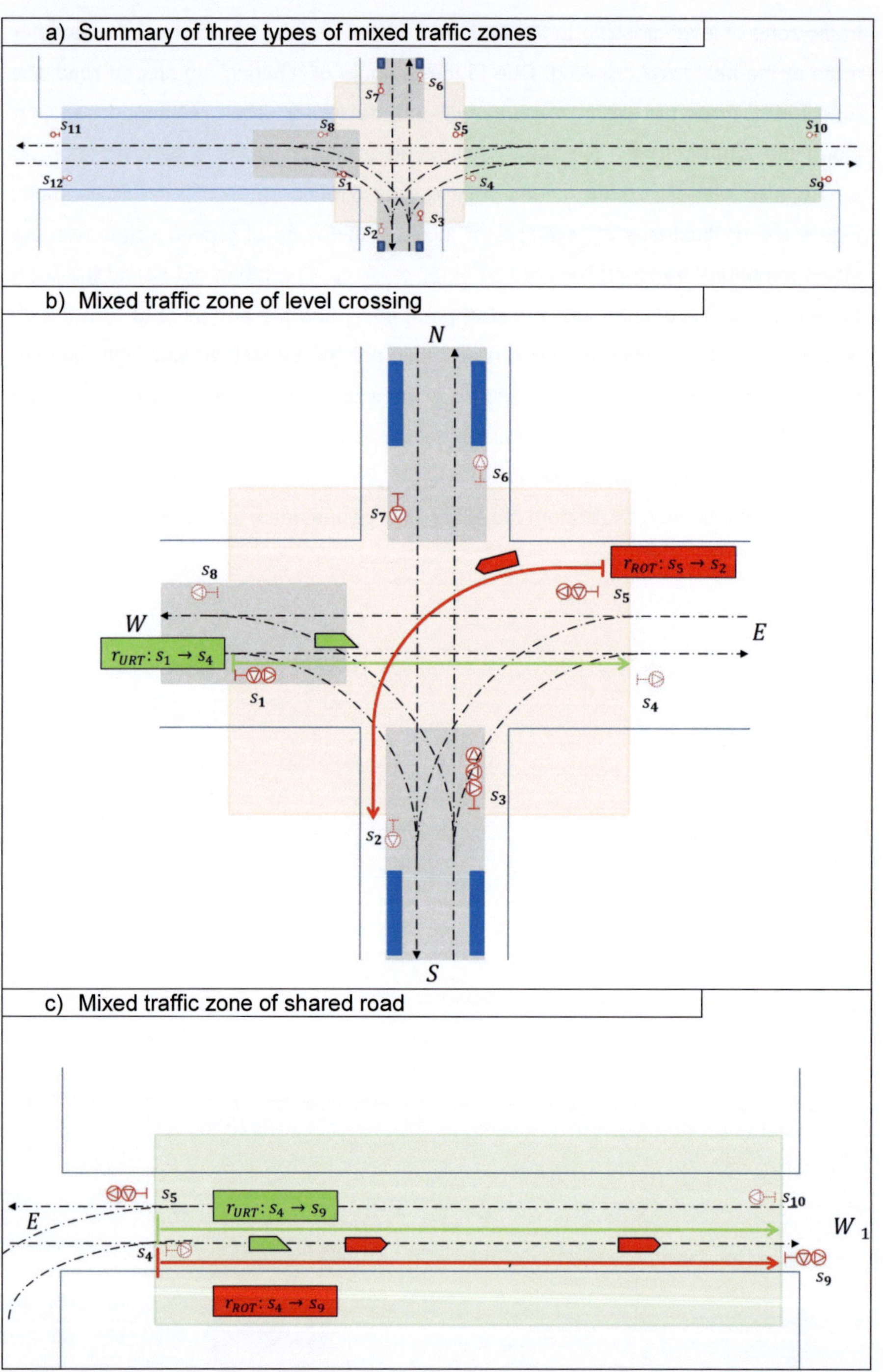

a) Summary of three types of mixed traffic zones
s11
s12
s8
s1
s7
s6
s5
s3
s2
s4
s10
s9
b) Mixed traffic zone of level crossing
N
s6
s7
s8
rROT: s5 → s2
s5
W
E
rURT: s1 → s4
s1
s4
s3
s2
S
c) Mixed traffic zone of shared road
s5
rURT: s4 → s9
s10
E
W
s4
s9
rROT: s4 → s9

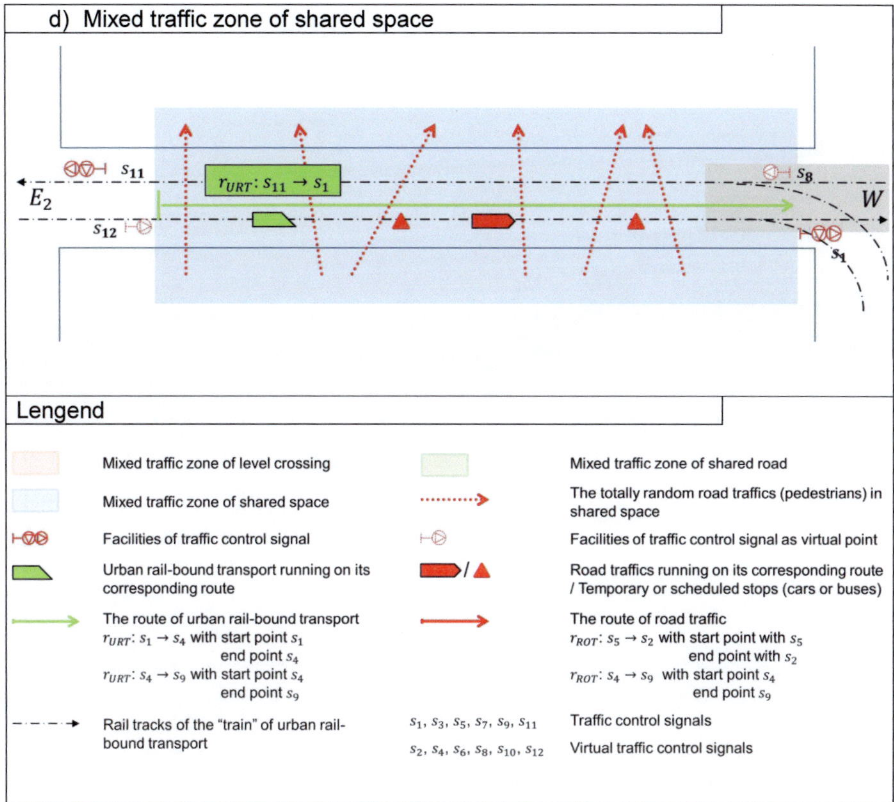

Figure 5-2: Microscopic Description of Routes of Urban Mixed Traffic at Mixed Traffic Zones (Source: developed based on [Martin & Liu 2016a])

As mentioned above, it is necessary to define the routes of urban rail-bound transport and road traffic (except in shared space) in this algorithm in order to model the event-driven system. The DFG-research project [Martin & Liu 2016a] described a locking table of route related locking of urban mixed traffic to declare the routes of urban mixed traffic and the relationships among them, in order to express explicitly the interactions among urban mixed traffic especially for mixed traffic zone of level crossing. For mixed traffic zone of shared road, urban rail-bound transport routes and road traffic routes interfere and coexist with each other. Comparably, the road traffic in mixed traffic zone of shared space may interfere with the urban rail-bound transport at any time. The locking table of route related locking of urban mixed traffic has to be defined as reference, which is shown in Table 5-1.

| | R_{ROT} | | | | R_{URT} | | | | | | | |
	$r_{ROT,1}$	$r_{ROT,2}$	$r_{ROT,3}$	$r_{ROT,4}$	$r_{URT,1}$	$r_{URT,2}$	$r_{URT,3}$	$r_{URT,4}$	$r_{URT,5}$	$r_{URT,6}$	$r_{URT,7}$	$r_{URT,8}$
$r_{ROT,1}$		1	1	1	1	1	1	1	1	1	0	0
$r_{ROT,2}$			1	1	1	1	1	1	0	0	1	1
$r_{ROT,3}$				1	1	1	1	1	1	1	0	0
$r_{ROT,4}$					1	1	1	1	0	0	1	1
$r_{URT,1}$						0	1	1	1	1	1	1
$r_{URT,2}$							1	1	0	1	0	1
$r_{URT,3}$								0	1	1	0	1
$r_{URT,4}$									0	1	1	1
$r_{URT,5}$										0	0	1
$r_{URT,6}$											1	1
$r_{URT,7}$												0
$r_{URT,8}$												

Table 5-1: The Locking Table of Route Related Locking for Urban Mixed Traffic[32] (Source: modified based on [Martin & Liu 2016b] & [Martin & Liu 2016a])

In the locking table of route related locking, it is possible to judge whether any two routes of urban mixed traffic in the investigated mixed traffic zone are compatible or exclusive. Two routes are conflicted, which means that they are exclusive to each other and that the two routes cannot pass through the mixed traffic zone at the same time. Therefore, "1" is marked in the locking table of route related locking. An example of two exclusive routes is shown in Figure 5-3 a). The road traffic route $r_{RT,1}$ (red arrow line) is in conflict with the urban rail-bound transport route $r_{URT,1}$ (green arrow line). They are mutually exclusive routes and are marked as "1" in the table.

---

[32]It shows the information of an example of mixed traffic zone of level crossing introduced in Appendix I: Basic Information of the Investigated Example.

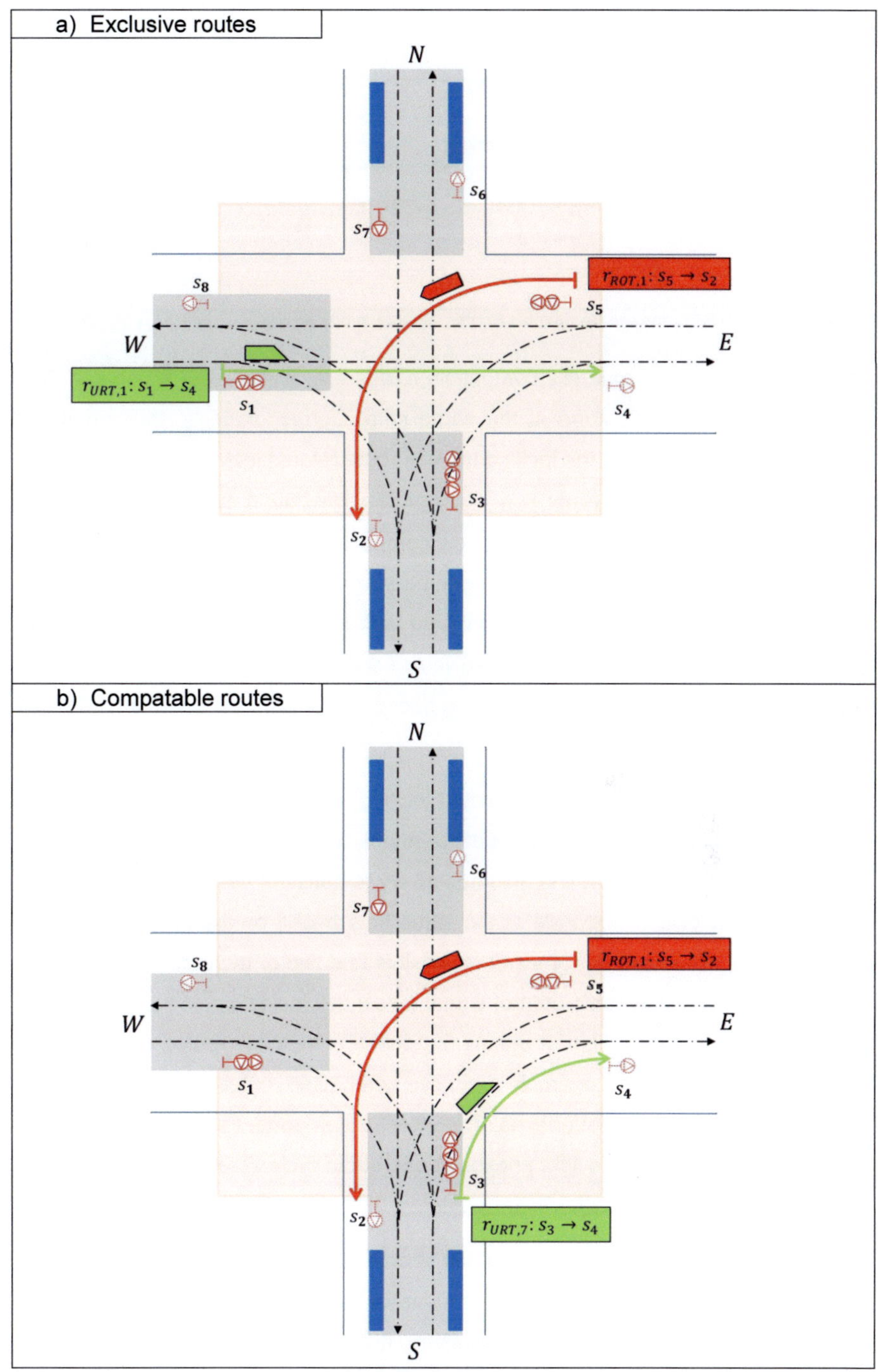

a) Exclusive routes
N
s_6
s_7
s_8
$r_{ROT,1}: s_5 \rightarrow s_2$
s_5
W
E
$r_{URT,1}: s_1 \rightarrow s_4$
s_1
s_4
s_3
s_2
S

b) Compatable routes
N
s_6
s_7
s_8
$r_{ROT,1}: s_5 \rightarrow s_2$
s_5
W
E
s_1
s_4
s_3
$r_{URT,7}: s_3 \rightarrow s_4$
s_2
S

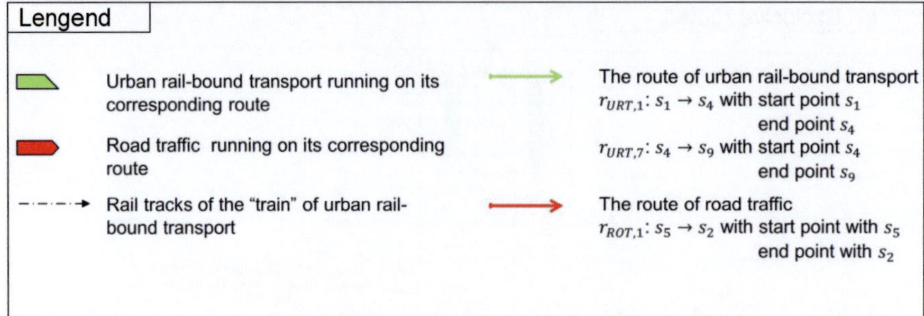

Figure 5-3: Exclusive Routes and Compatible Routes (Source: modified based on [Martin & Liu 2016a])

On the other hand, the mutually compatible routes are conflict-free, which can pass through the mixed traffic zone at the same time. Correspondingly, the compatible routes are marked as "0" in the locking table of route related locking. There is no conflict between road traffic route $r_{ROT,1}$ and urban rail-bound transport route $r_{URT,7}$ shown in Figure 5-3 b), which are mutually compatible routes and are marked in the table as "0". As a reference, it is very useful during the modeling of the event-driven system to have the probability of urban mixed traffic moving simultaneously on theirs corresponding routes in an investigated mixed traffic zone according to the determined locking table of route related locking based on the practical operation [Martin & Liu 2016a].

Based on the locking table of route related locking for urban mixed traffic, the possible simultaneous movements of compatible routes are explicit for mixed traffic zone of level crossing. Meanwhile, it is also necessary to consider whether these compatible routes are allowed to operate at the same time based on the operation rules of corresponding traffic control signals. It is possible that two or more compatible routes controlled by the same traffic control signal follow the indication of the same traffic light phase in one traffic cycle. It means that these compatible routes can share the same green light phase for example. Or these compatible routes are separately controlled by different traffic light phases. Therefore, it is important and necessary to determine the rotation of traffic light phases in one traffic cycle at level crossing to complete the locking table of route related locking.

The normal rotation of traffic light phases should be determined with the initial sequence and the corresponding time durations of various traffic light phases in one traffic cycle at the investigated level crossing. It can reflect the movements of road traffic without any urban rail-bound transport. The **initial sequence** of the normal rotation of traffic light phases (shown with green light phase $GP_{r_{ROT,p}}$ for road traffic route $r_{ROT,p}$) can be set as example based on the locking table of route related locking in Table 5-1 with four exclusive road traffic routes.

$$GP_{r_{ROT,1}} \rightarrow GP_{r_{ROT,2}} \rightarrow GP_{r_{ROT,3}} \rightarrow GP_{r_{ROT,4}} \rightarrow GP_{r_{ROT,1}} \rightarrow GP_{r_{ROT,2}} \rightarrow \cdots$$

Furthermore, the corresponding time durations of various traffic light phases of a road traffic route in one traffic cycle controlled by the corresponding traffic control signal at the investigated level crossing can be defined as the **effective occupation time** of a road traffic route ($o_{ROT,p}$) in a traffic cycle. [Martin & Liu 2016a] gave the definition of the effective occupation time ($o_{ROT,p}$). It is the time duration of a road traffic route ($r_{ROT,p}$) starting from its entry till the time point that other urban mixed traffic on theirs routes can enter the investigated mixed traffic zone or from an all red light phase[33]. As shown in Figure 5-4 b), the effective occupation time of road traffic route ($o_{ROT,p}$) can be determined by the time durations of parts of different phases of road traffic route in one traffic cycle, which is an important input of the model in this algorithm.

The effective occupation time of road traffic route ($o_{ROT,p}$) refers to the time duration of a road traffic route ($r_{ROT,p}$)[34] that occupies the mixed traffic zone of level crossing. The starting point of the effective occupation time of road traffic route ($o_{ROT,p}$) is the starting time of green light phase ($GP_{ROT,p}$ in seconds) for the road traffic route $r_{ROT,p}$. The yellow light phase following its green light phase ($YP(G)_{ROT,p}$) also forbids other exclusive routes to pass and should be calculated in the effective occupation time of the road traffic route ($o_{ROT,p}$).

---

[33] All red light phase ($RP(A)$): is the possible time duration for indication of all traffic lights in red, which is the time duration between successive effective occupation time of two exclusive routes in a traffic cycle.

[34] The road traffic route with the subscript "p or $p + 1$" here shows that the green light phase of road traffic route $r_{ROT,p+1}$ is the next green light phase after completing the effective occupation time of road traffic route $r_{ROT,p}$. It is an example sequence of green light phase of road traffic routes in a traffic cycle.

In practice, after the yellow light phase changed from its green light phase $(YP(G)_{ROT,p})$, this traffic light phase changes to the red light phase. However, for security reason, at this time point it is not allowed to start the next green light phase for another road traffic route $(GP_{ROT,p+1})$ in the defined normal rotation at once. Certain time duration is necessary to ensure that the level crossing is clear. All road traffic on the road traffic route $(r_{ROT,p})$ that occupied the level crossing during the previous green light phase $(GP_{ROT,p})$ and the yellow light phase $(YP(R)_{ROT,p})$ should clear the level crossing. This time duration is also included in the effective occupation time of the road traffic route $(o_{ROT,p})$. The certain part time duration of red light phase of the road traffic $(RP_{ROT,p})$ starts from the end of the yellow light phase $(YP(G)_{ROT,p})$ until the starting point of the next green light phase $(GP_{ROT,p+1})$ of the following road traffic route $(r_{ROT,p+1})$ in the defined normal rotation, which is equal to the yellow light phase changed from red light phase of the next road traffic route $(YP(R)_{ROT,p+1})$.

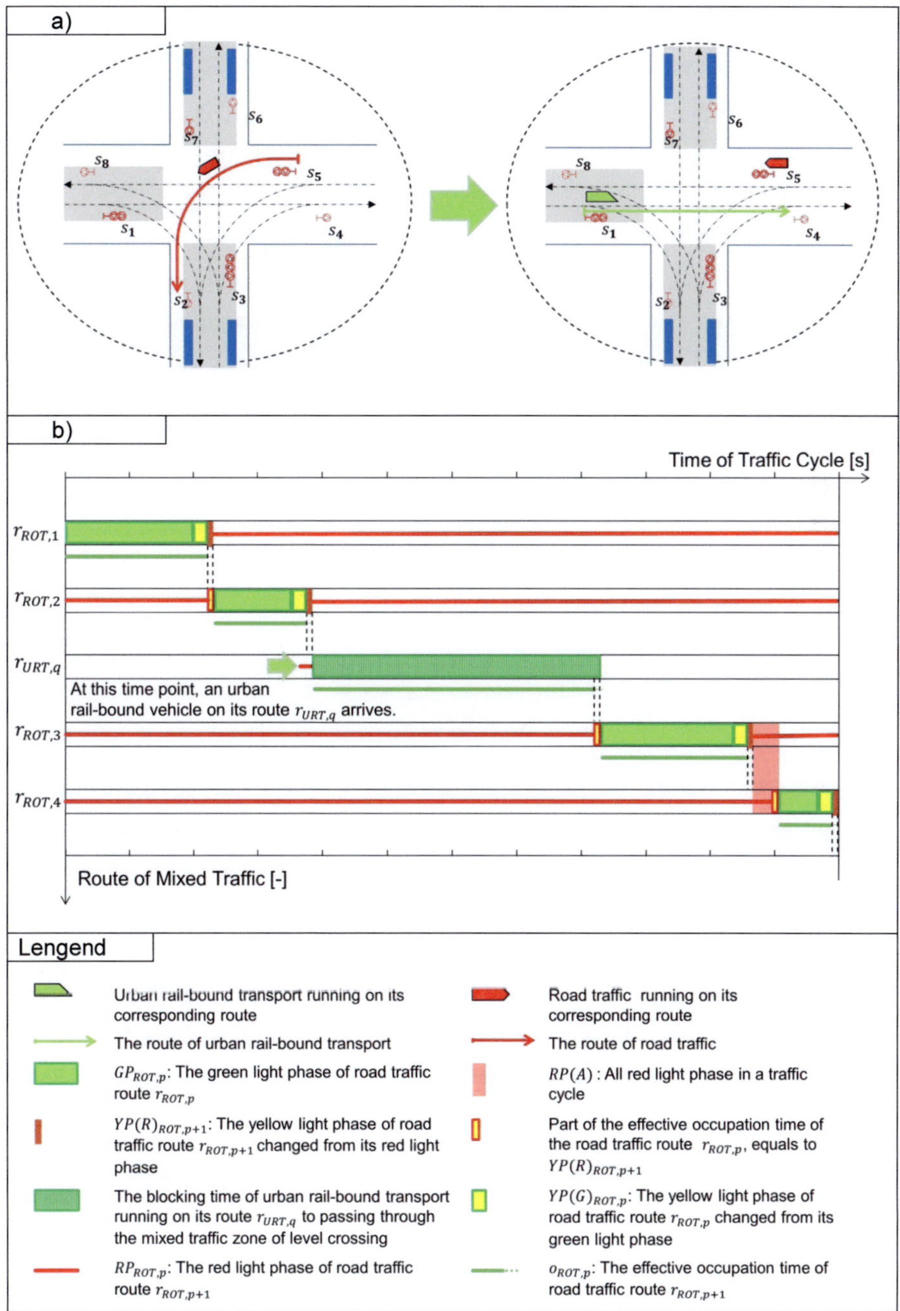

Figure 5-4: The Effective Occupation Time of Road Traffic Route with Different Traffic Light Phases of Traffic Control Signal (Source: modified based on [Martin & Liu 2016a])

It is also possible that the effective occupation time of the road traffic route ($o_{ROT,p}$) is followed by an all red light phase ($RP(A)$) when all phases are red. In this algorithm, all red light phase $RP(A)$ is considered as a vacuous time period that road traffic cannot occupy the mixed traffic zone. When there is urban rail-bound transport arriving during this time duration, it can directly enter and pass through the mixed traffic zone without road traffic influences.

In conclusion, the effective occupation time of road traffic route ($o_{ROT,p}$) can be derived based on [Martin & Liu 2016a] in the normal rotation without urban rail-bound transport.

$$o_{ROT,p} = GP_{ROT,p} + YP(G)_{ROT,p} + YP(R)_{ROT,p+1} \qquad (\text{ 5-4 })$$

With:

$GP_{ROT,p}$ The green light phase of road traffic route $r_{ROT,p}$

$YP(G)_{ROT,p}$ The yellow light phase changed from green light phase of road traffic route $r_{ROT,p}$

$YP(R)_{ROT,p+1}$ The yellow light phase changed from red light phase of the next road traffic route $r_{ROT,p+1}$

p The index of the road traffic routes, $p \in [1, m]$

In some situations, the yellow light phases $YP(G)_{ROT,p}$ and $YP(R)_{ROT,p+1}$ may have overlap, and the effective occupation time of road traffic route should be adjusted based on the requirements in practice. For the specific operation process in the simulation model of event-driven system with this algorithm, the effective occupation time in each traffic cycle is calculated based on the simulated traffic light phases instead of the pre-defined traffic light phases in normal rotation. Consequently, the corresponding time duration of **traffic cycle (TC)** can be determined by the sum of all the traffic light phases in one traffic cycle of the normal rotation at the investigated level crossing.

In addition, if an urban rail-bound transport arrives at the investigated level crossing as described in Subchapter 2.1, the normal rotation of road traffic routes will be interrupted, and this can occur at any random time point during the investigated time period. The ongoing road traffic on road traffic routes have to complete their minimum green light phase ($(min)GP_{ROT,p}$) before giving the way to the urban rail-bound transport in some cases (see Subchapter 5.3.2). Therefore, the minimum effective occupation time $(min)o_{ROT,p}$ of road traffic route $r_{ROT,p}$ can be calculated. Also in some situations, the yellow light phase may be adjusted and can be set in the modeling of the algorithm based on the actual situation and requirements.

$$(min)o_{ROT,p} = (min)GP_{ROT,p} + YP(G)_{ROT,p} + YP(R)_{ROT,p+1} \qquad (\,5\text{-}5\,)$$

With:

$\qquad (min)GP_{ROT,p}$ The minimum green light phase of road traffic route $r_{ROT,p}$

As a possible case shown in Figure 5-4, the urban rail-bound transport $r_{URT,q}$ is arriving while the road traffic route $r_{ROT,2}$ is still on its yellow light phase $(YP(G)_{ROT,2})$. Because the two routes are exclusive in this example, the urban rail-bound transport has to wait for a while until the road traffic on the route $r_{ROT,2}$ completes their occupation time of road traffic route $o_{ROT,2}$. For security reason, the road traffic route $r_{ROT,2}$ has to block the investigated level crossing for the time duration of $YP(R)_{ROT,3}$ to make sure all road traffic clear, even though there is no "yellow light phase" for urban rail-bound transport before its progression [Martin & Liu 2016a]. For urban rail-bound transport, the time duration to pass through the investigated level crossing is defined as the **blocking time (BL_{URT})** of the level crossing. After that, the road traffic route goes back to the defined normal rotation, which starts with its green light phase $(GP_{ROT,3})$. The rotation and duration of phases of the traffic control signals for road traffic are updated cyclically by pre-defined rules.

For mixed traffic zone of shared road, the above inputs for level crossing are also needed because the traffic control signals at the next adjacent level crossing in the direction indirectly control the movements of road traffic on routes, and further influence the operation of urban rail-bound transport. Therefore, the time duration of **traffic cycle (TC)** and the corresponding **traffic light phases**[35] in the direction of the route on the investigated shared road have to be given as inputs for modeling of event-driven system for shared road.

---

[35] The traffic light phases here indicate that they can control the movements of urban mixed traffic on theirs routes at the end of the investigated shared road.

In addition, because of the different speeds of urban rail-bound transport and road traffic, the urban rail-bound transport is hindered by road traffic. The urban rail-bound transport has to follow the road traffic movements. Therefore, it is necessary to model the **proportion of various road traffic types** with the given corresponding **speeds of road traffic (V_{ROT})** to determine the running time of urban rail-bound transport. Meanwhile, the **running distance ($s_{URT,i}$)** with direction related to the corresponding **running time ($t_{URT,i}$)**[36] of urban rail-bound transport (U_i) before it is hindered by road traffic, which can be determined based on the dynamics of urban rail-bound transport. Accordingly, the **length of the investigated shared road (S)** should be prepared.

Comparably, in mixed traffic zone of shared space, urban rail-bound transport may be influenced by road traffic (pedestrians, or stops of cars and buses) without the control of traffic control signals. Pedestrians move totally randomly and cars with low traffic load also stop temporarily at random. Moreover, scheduled bus stops are set as fixed point with information of time and location. Therefore, it is unnecessary to define the inputs related to the traffic control signals. However, once the urban rail-bound transport is hindered along the shared space, deceleration and stop will happen and cause extra waiting time for urban rail-bound transport. In this developed algorithm, this course is assumed as an average value of **extra waiting time (WT_{URT})** to represent the average influences caused by road traffic in shared space. It means that each time an urban rail-bound transport encounters hindrance by pedestrians or stops of cars and buses along the shared space, there is extra waiting time (WT_{URT}) arising. It happens at every time of hindering, which is an iterative process. The urban rail-bound transport has to restart and accelerate based on the dynamics of urban rail-bound transport. The **running distance ($s_{URT,i}$)** with direction related to the corresponding **running time ($t_{URT,i}$)**[37] of urban rail-bound transport (U_i) before the

---

[36] The corresponding running time ($t_{URT,i}$) here indicates the time course (in seconds) of the urban rail-bound transport (U_i) during the investigated time period with the running distance ($s_{URT,i}$) in the coordinate ($x_{URT,i}$, 0) along the investigated shared road without road traffic influences yet.

[37] The corresponding running time ($t_{URT,i}$) here in mixed traffic zone of shared space also indicates the time course (in seconds) of the urban rail-bound transport (U_i) from the end of a hindrance to the next hindrance by road traffic during the investigated time period. It will be updated in each new iterative with the total time course of all occurred iterative and extra waiting time.

hindrances in each iterative can be derived as the basis of this algorithm for the model of an event-driven system for shared space.

5.2.3   Relevant Rules of Traffic Control Signal

Road traffic follow the instruction of traffic control signals at mixed traffic zone of level crossing, and indirectly in mixed traffic zone of shared road. Therefore, the influences of road traffic are represented by the rotation and time duration of traffic light phases in a traffic cycle during an investigated time period. It is necessary to set the basic relevant rules traffic control signaling system in this algorithm. In the DFG-research project [Martin & Liu 2016a], three main operational rules on principle were discussed.

- Sequential rule
- Application rule
- Limitations of traffic light phase rule

In this dissertation, another rule is further explored in depth.

- Route set rule

a) **Sequential rule** for traffic control signaling system

The sequence of traffic light rotation commands the sequence of road traffic on corresponding routes to proceed in the indicated direction in any traffic cycle.

i.   Ordering principle

The traffic light rotation is based on the pre-defined sequence, which means the road traffic on corresponding routes have to follow the pre-defined sequence kept in each traffic cycle. Even though, it can be interrupted at any random time point by an arriving urban rail-bound transport, the sequence has to be maintained in the following operation.

ii.   Disordering principle

The traffic light phases can rotate and should be flexible depending on the requirements, which can be changed in the operation process. When there is an urban rail-bound transport that interrupts the normal rotation, the road traffic routes that are compatible with this urban rail-bound transport, meanwhile, can still proceed theirs green light phase without restriction of the original sequence.

b) **Application rule** for urban rail-bound transport

As described in Subchapter 2.1, the urban rail-bound transport has the right to apply for the priority to pass through the level crossing through the device along the rail tracks in advance. [Martin & Liu 2016a] assumed that the applying point is defined as the start point of the urban rail-bound transport route asking for entering and occupying the mixed traffic zone of level crossing. It is usually not far away from the facility of traffic control signal in practice. At the time point of urban rail-bound transport arrival, the interactions between road traffic and urban rail-bound transport are described in various cases in the event-driven system (detailed in Subchapter 5.3).

c) **Limitation of traffic light phase rule** for traffic control signaling system

There are two regulations for traffic control signals that restrict the absolute priority of urban rail-bound transport at mixed traffic zone of level crossing.

- The maximum red light phase for road traffic
- The minimum green light phase for road traffic

It is necessary to fulfill in this algorithm as prerequisites preferentially. According to the requirements in reality, the values of these two prerequisites have to be given as inputs for event-driven system in this algorithm.

d) **Route set rule** for road traffic routes

Route set is defined as one set of some compatible road traffic routes at a mixed traffic zone of level crossing. The road traffic routes in one route set are conflict-free and can move through the level crossing under green light phase at the same time[38]. In Figure 5-5 a), the road traffic routes $r_{ROT}: s_5 \rightarrow s_8$ and $r_{ROT}: s_1 \rightarrow s_4$ are in one route set of road traffic for instance.

$$R_{ROT,set_k} = \left\{ r_{ROT,f_k(1)}, r_{ROT,f_k(2)}, \cdots, r_{ROT,f_k(p')}, \cdots, r_{ROT,f_k(m_{set_k})} \right\} \qquad (5\text{-}6)$$

$$R_{ROT} = \bigcup_{r_{ROT,f_k(p')} \in R_{ROT}} R_{ROT,set_k} \qquad (5\text{-}7)$$

With:

---

[38] Compatible routes in one route set include the routes under the same traffic light phase of the same traffic control signal or under the same traffic light phase of different traffic light control signals.

R_{ROT,set_k} — The route set of road traffic routes at a mixed traffic zone with $R_{ROT,set_k} \subseteq R_{ROT}$

k — The index of the route set of road traffic routes, $k \in [1, K]$

K — The number of route sets in a traffic cycle

$r_{ROT,f_k(p')}$ — The road traffic route in route set k with $f_k(i)$: $\{1, 2, \cdots, p', \cdots, m_{set_k}\} \to \{1, 2, \cdots, p, \cdots, m\}$

p' — The index of the compatible road traffic routes in the route set k, $p' \in \left[1, m_{set_k}\right]$

m_{set_k} — The number of elements (routes of road traffic) of the route set k with $\left|R_{RT,set_k}\right| = m_{set_k}$

If all routes of road traffic operating at an investigated level crossing are compatible with one another, there is only one route set including all road traffic routes as elements ($m_{set_k} = m$ and $R_{RTO,set_k} = R_{ROT}$). On the other hand, if all routes of road traffic are independent and exclusive to one another, each route set has only one element of road traffic route ($m_{set_k} = 1$ and $R_{ROT,set_k} = \{r_{ROT,p}\}$). The number of route sets can be seen as the number of routes m.

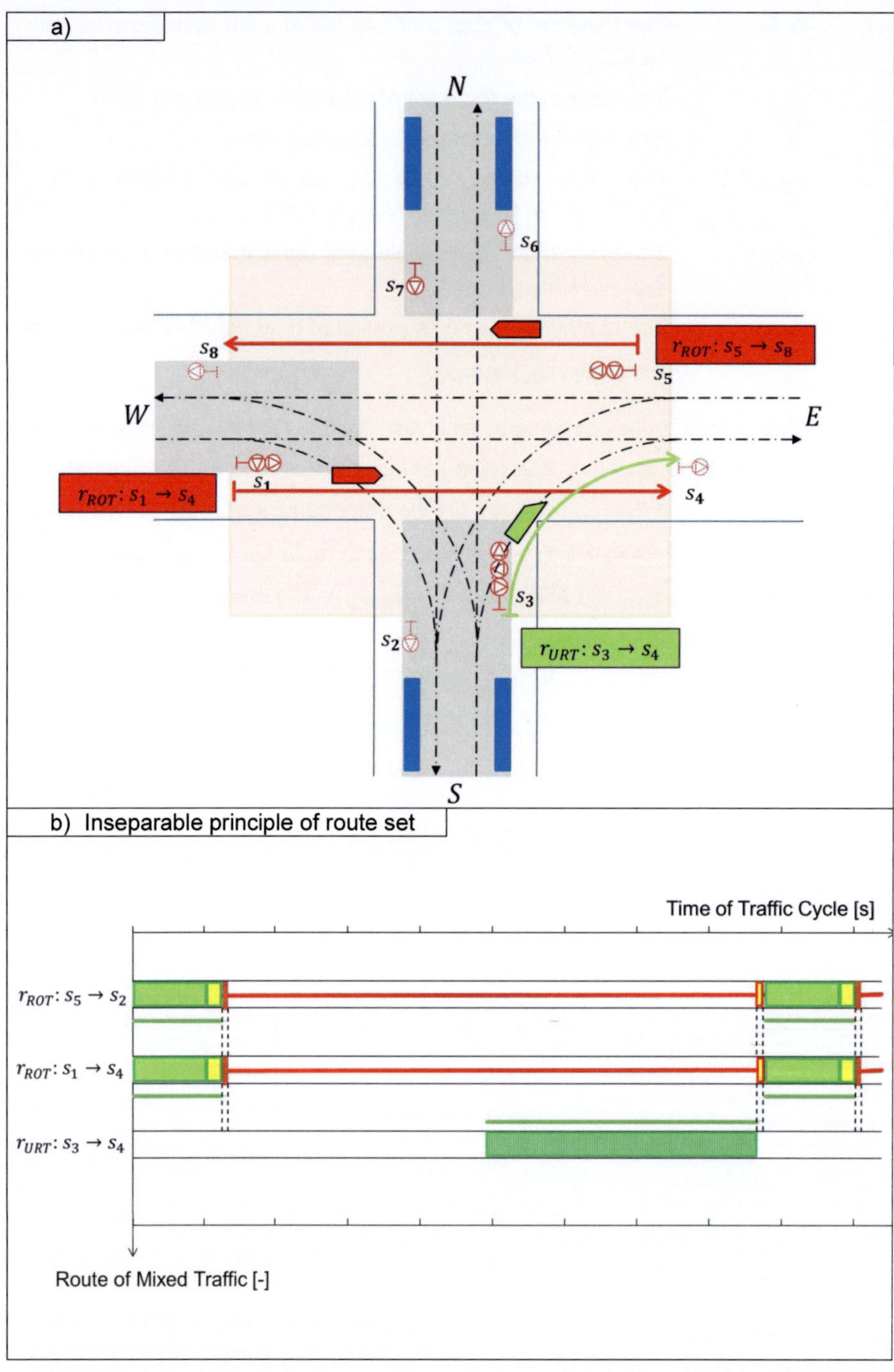
a)
N
s_6
s_7
s_8
$r_{ROT}: s_5 \rightarrow s_8$
s_5
W
E
$r_{ROT}: s_1 \rightarrow s_4$
s_1
s_4
s_3
$r_{URT}: s_3 \rightarrow s_4$
s_2
S
b) Inseparable principle of route set
Time of Traffic Cycle [s]
$r_{ROT}: s_5 \rightarrow s_2$
$r_{ROT}: s_1 \rightarrow s_4$
$r_{URT}: s_3 \rightarrow s_4$
Route of Mixed Traffic [-]

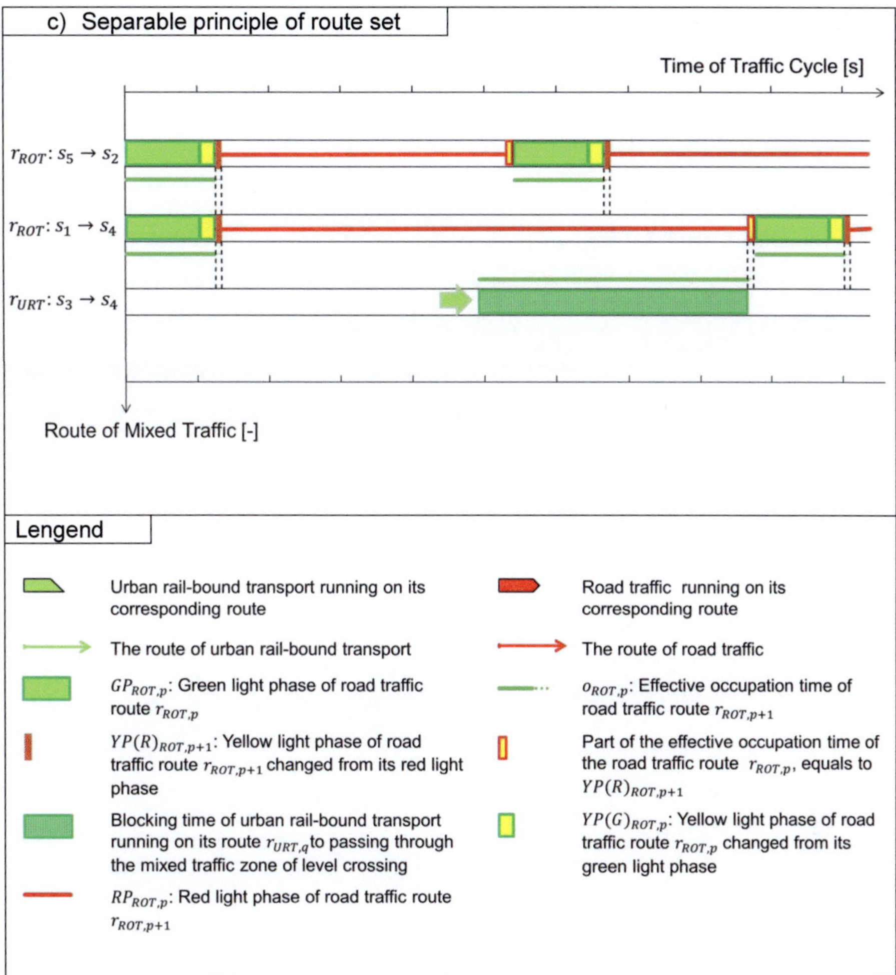

Figure 5-5: Route Set of Road Traffic

For various investigated mixed traffic zones of level crossing, the route sets of road traffic can be ruled differently under different requirements as illustrated in Figure 5-5 b) and c) respectively:

i.   Inseparable principle of route set

The route sets have to be kept all along in the operation process under the inseparable principle of route sets. It means that all compatible routes in one route set are bound together and cannot move separately as shown in Figure 5-5 b). At mixed traffic zone of level crossing, the traffic control signals controlling this route set with these road traffic routes also have the same traffic light phase. It can be green light phase to allow road traffic to proceed in corresponding directions as denoted; or it can be yellow light phase to warn traffic that the signal is about to change to red light or indicates that the signal is about to change to green light; or it can be red light phase to prevent traffic from passing the level crossing.

For one defined route set R_{ROT,set_k}, if there is an urban rail-bound transport (U_i) on its route proceeding, some routes in this route set are compatible with this urban rail-bound transport route (CR_{ROT,set_k}), while other routes in this route set are exclusive with the urban rail-bound transport route (ER_{ROT,set_k}). For each interacted urban rail-bound transport, the route set R_{RTO,set_k} can be defined as a union of the two subsets (CR_{ROT,set_k} and ER_{ROT,set_k}). Each of the two route sets includes part of the elements of original route set R_{ROT,set_k}. Therefore, the sum of numbers of the two route sets is m_{set_k}.

$$R_{ROT,set_k} = CR_{set_k,U_i} \cup ER_{set_k,U_i} \qquad (5\text{-}8)$$

$$
\begin{aligned}
& CR_{set_k,U_i} \cup ER_{set_k,U_i} \\
& := \left\{ r_{ROT,f_k(p')} \mid \left(r_{ROT,f_k(p')} \in CR_{set_k,U_i} \right) \vee \left(r_{ROT,f_k(p')} \in ER_{set_k,U_i} \right) \right\}
\end{aligned}
\qquad (5\text{-}9)
$$

With:

U_i — The urban rail-bound transport U_i that is running on the route $r_{URT,q}$ in mixed traffic zone

CR_{set_k,U_i} — The subset with the elements of road traffic routes that are compatible with the arriving urban rail-bound transport U_i belongs to route set R_{RTO,set_k}, $CR_{set_k,U_i} \subseteq R_{ROT,set_k}$

ER_{set_k,U_i} — The subset with the elements of road traffic routes that are exclusive to the arriving urban rail-bound transport U_i belongs to route set R_{ROT,set_k}, $ER_{set_k,U_i} \subseteq R_{ROT,set_k}$

Because of the restriction of "Inseparable principle", neither routes in original route set R_{ROT,set_k} exclusive with this urban rail-bound transport route ($r_{ROT,f_k(p')} \in ER_{set_k,U_i}$) nor routes compatible with this urban rail-bound transport route ($r_{ROT,f_k(p')} \in CR_{set_k,U_i}$) can be given green light phase to proceed with urban rail-bound transport U_i. All the road traffic routes in one route set R_{ROT,set_k} have to operate together all the time. To sum up, the road traffic routes $r_{ROT,f_k(p')}$ in one route set R_{ROT,set_k} can be treated as one "route".

ii.   Separable principle of route set

Under this condition, compared with the "Inseparable principle", the routes of road traffic in one route set R_{ROT,set_k} can move together or separately based on the practical requirements. Theoretically, the routes belonging to one route set can operate together. However, in some situations, as shown in Figure 5-5 c), some routes of road traffic in a route set are compatible with the arriving urban rail-bound transport. It is allowed that those road traffic routes can proceed together with the urban rail-bound transport. The subset CR_{ROT,U_i} in the route set of road traffic R_{ROT,set_k} can be separated from the route set to move together under the same indication of traffic light phase.

For a mixed traffic zone of level crossing, the route sets can be defined variously. Therefore, for a specific investigated level crossing it is necessary to determine the required route sets integrated with the pre-defined locking table of route related locking (see Table 5-1). Moreover, the corresponding rules of sequence and route sets described above for the traffic control signals at a mixed traffic zone of level crossing have to be set at the beginning to guide the interactions of various cases in the modeled event-driven system (see Subchapter 5.3).

5.3   Event-driven Cases

In this algorithm, event-driven system is utilized to build the model of random interactions between urban rail-bound transport and road traffic in different types of mixed traffic zones of level crossing, shared road and shared space respectively in this subchapter. The basic principle of this algorithm was discussed in the DFG-research project [Martin & Liu 2016a]. The trigger of predicted interactions between urban rail-bound transport and road traffic is defined in the form of "event", which is realized in various cases. The randomness of an occurrence can be reflected by the event through possible changes of status of the urban mixed traffic at a specific time point during the investigated time period [Martin & Liu 2016a].

5.3.1 Overview

Based on the developed algorithm, an event is triggered in the model as described in the DFG-research project [Martin & Liu 2016a], when an urban rail-bound vehicle (U_i) on its corresponding route ($r_{URT,q}$) arriving at the investigated mixed traffic zone of level crossing at a random specific time point during the investigated time period. For mixed traffic zone of shared road and shared space, it is similar that an event is triggered when an urban rail-bound vehicle (U_i) on its corresponding route ($r_{URT,q}$) is hindered by road traffic at random time points during the investigated time period at the investigated mixed traffic zone. When an event is triggered, the corresponding responses (i.e. the predicted interactions) of the urban rail-bound transport and road traffic are defined in various cases based on the pre-determined rules and inputs (described in Subchapter 5.2).

Figure 5-6 shows the workflow of this developed algorithm with the event-driven system in the model, which is the basic framework of this algorithm in studying mixed traffic zones of level crossing, shared road and shared space. The randomly occurred cases triggered by events are selected by the assessment and judgment of the specific situation at the specific time point of the event.

There are four steps to implement the model with the event-driven system when an event is triggered in various investigated mixed traffic zones, which were described in [Martin & Liu 2016a] as follows:

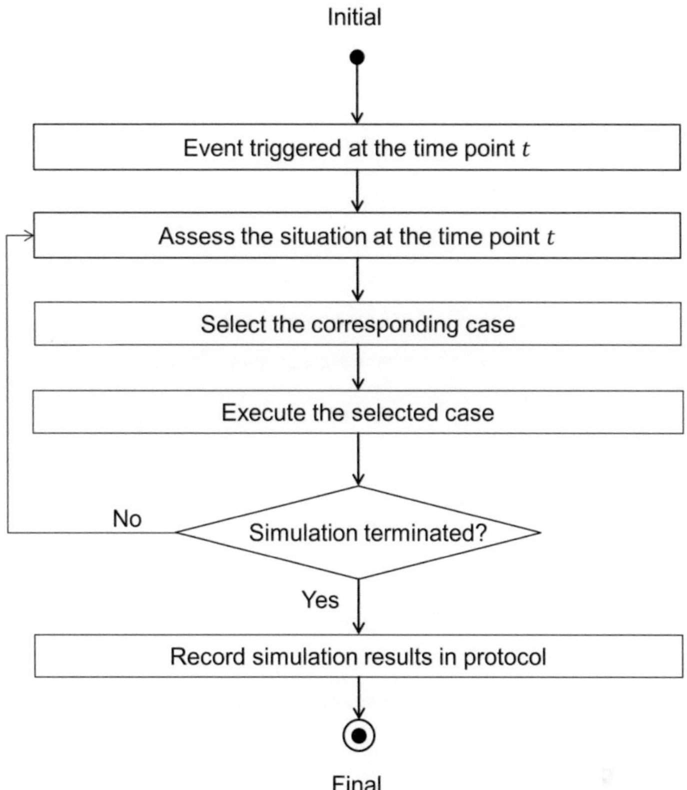

Figure 5-6: The Workflow of Event-Driven Simulation with the Developed Algorithm (Source: modified based on [Martin & Liu 2016a])

Step 1: Assess the situation at the time point t

In order to determine the operation status changes of the urban mixed traffic due to the triggered event at the time point t during the investigated time period, it is necessary to assess the temporal situation that is the status of the interaction between urban rail-bound transport and road traffic.

Step 2: Select the corresponding case

The corresponding case representing the specific responses (i.e. changes of status) of urban rail-bound transport and road traffic is selected based on the assessment of the interaction status at the time point t in step1.

Step 3: Execute the selected case

In the simulation model, the selected case is executed to implement the specific responses, which is to materialize the respective status changes of urban mixed traffic caused by the triggered event.

Step 4: Judge whether the simulation is terminated?

After the executed case is completed, this one iterative of event is over in the simulation process. However, it is possible that the investigated time period is still in progress. The simulation continues to proceed to next iterative of event until the investigated time period is completed, which means that the simulation is terminated. The results of simulation are recorded in the output protocol consequently.

For various types of mixed traffic zones, the corresponding cases that reflect the responses of urban mixed traffic when an event is triggered vary. Figure 5-7 shows the general judgments for cases at various investigated mixed traffic zones and the corresponding results, which will be interpreted further in the following subchapters. According to the input and relevant rules described in Subchapter 5.2, six cases will be discussed in next Subchapter 5.3.2 under various statuses of interactions between urban rail-bound transport and road traffic. In next Subchapters 5.3.3 and 5.3.4, the mixed traffic zones of shared road and shared space are respectively summarized in one logic process to represent the status changes of urban rail-bound transport. Consequently, based on the recorded simulation results the extra waiting time of urban rail-bound transport can be derived by the defined indicators calculated in Subchapter 5.4.

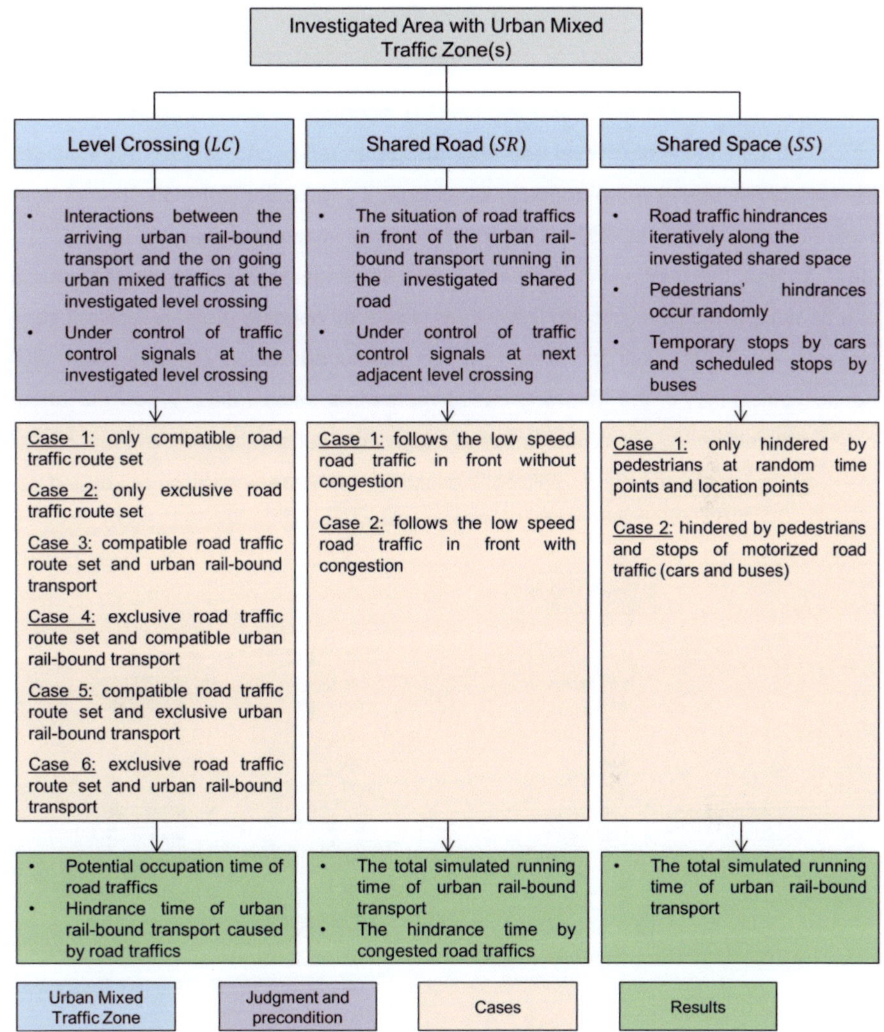

Figure 5-7: Overview of the Algorithm for Various Mixed Traffic Zones

5.3.2   *LC* Cases for Level Crossing

At mixed traffic zone of level crossing, the responses of road traffic vary due to the competition and interruption of urban rail-bound transport, leading to different impacts on the urban rail-bound transport. Various cases are further developed and discussed based on the introduction in [Martin & Liu 2016a] at a random specific time point when an event is triggered. Figure 5-8 illustrates six cases under different event trigger conditions with the different occupation statuses of road traffic at level crossing, when an urban rail-bound vehicle U_i on its route $r_{URT,q}$ arrives at an investigated level crossing at the specific time point t during the investigated time period.

If there is other urban rail-bound vehicle U_{i+1} when the urban rail-bound vehicle U_i arrives at the investigated level crossing, the urban rail-bound vehicle U_{i+1} indicates the compatible one or more trains. In simulation, it is possible that more than one urban rail-bound vehicle operate at the level crossing. When the urban rail-bound vehicle U_i arrives, the relationship between it and the ongoing urban rail-bound vehicle can be compatible or exclusive. If it is compatible, all ongoing urban rail-bound vehicles are compatible with the arriving urban rail-bound vehicle U_i. The urban rail-bound vehicle U_{i+1} in the following cases represents these trains. When the urban rail-bound vehicle U_{i+1} is exclusive to the urban rail-bound vehicle U_i, at least one of the ongoing urban rail-bound vehicles is exclusive to the arrived urban rail-bound vehicle U_i.

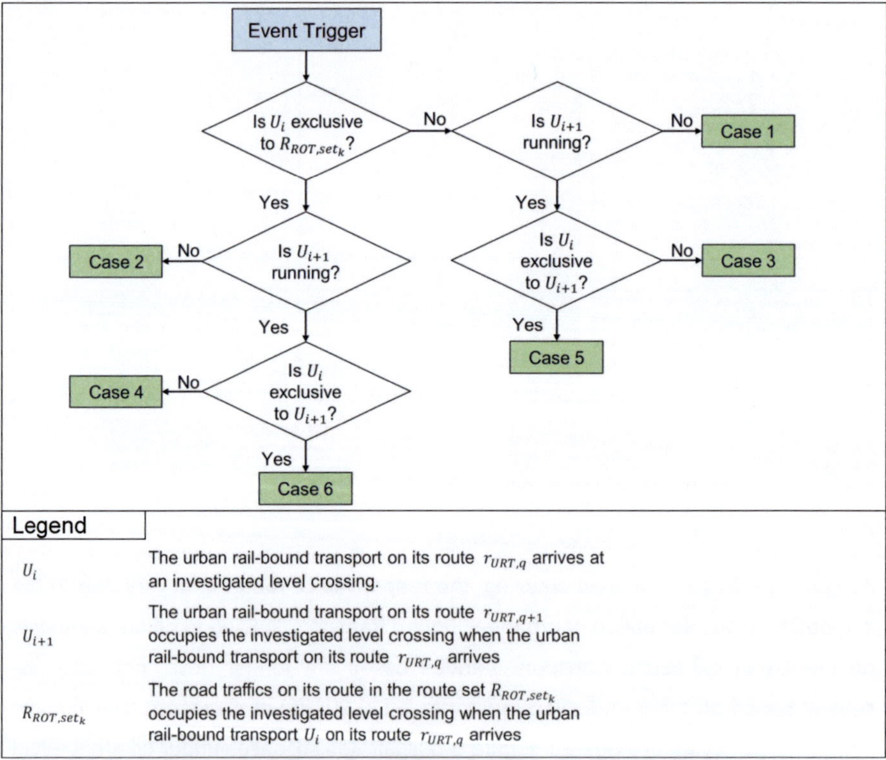

Legend	
U_i	The urban rail-bound transport on its route $r_{URT,q}$ arrives at an investigated level crossing.
U_{i+1}	The urban rail-bound transport on its route $r_{URT,q+1}$ occupies the investigated level crossing when the urban rail-bound transport on its route $r_{URT,q}$ arrives
R_{ROT,set_k}	The road traffics on its route in the route set R_{ROT,set_k} occupies the investigated level crossing when the urban rail-bound transport U_i on its route $r_{URT,q}$ arrives

Figure 5-8: Six *LC* Cases under Different Event Trigger Conditions for Mixed Traffic Zone of Level Crossing (Source: modified based on [Martin & Liu 2016a])

LC Case 1: Road traffic routes in route set R_{ROT,set_k} are compatible, and no other urban rail-bound transport is in progress

The event trigger condition for case 1 is that the ongoing road traffic route set R_{ROT,set_k} is compatible with urban rail-bound vehicle U_i at the investigated level crossing and that no other urban rail-bound vehicle operates at the time point when the urban rail-bound vehicle U_i arrives. As shown in Figure 5-9 and the comprehensive logic in Appendix Figure 4, the urban rail-bound vehicle can directly enter and occupy the mixed traffic zone of level crossing immediately with three possible outcomes depending on whether road traffic route set $R_{ROT,set_{RP}}$[39] exceeds its maximum red light phase.

a) When the urban rail-bound vehicle U_i proceeds to enter the investigated level crossing, it is the only ongoing road traffic route set R_{ROT,set_k}, no road traffic routes exceed maximum red light phase (Figure 5-9). Under this condition, it is possible that the surplus time duration of the effective occupation time of the road traffic route set $(Sur)o_{ROT,set_k}$ lasts longer than the blocking time of the arriving urban rail-bound vehicle BL_{U_i} to pass through the level crossing. There are no influences on urban rail-bound vehicle U_i or interruptions on road traffic with normal rotation of traffic control signals.

Otherwise, the blocking time of the arriving urban rail-bound vehicle BL_{U_i} is longer than the surplus time duration of the effective occupation time of the road traffic route $(Sur)o_{ROT,p}$. It is possible that other route set $R_{ROT,set_{k+1}}$[40] is compatible with urban rail-bound vehicle U_i after the occupation of road traffic route set R_{ROT,set_k}.

---

[39] The road route set $R_{ROT,set_{RP}}$ here indicates the road traffic route(s) in this route set that exceed the limitation of maximum red light phase. They have the same minimum green light phase to pass through the level crossing without separation. There can also be other routes (in corresponding route set) exceeding the maximum red light phase. In this algorithm, the decision is based on their sequence of exceeding the maximum red light phase.

[40] The route sets with the subscript "$k, k+1$" represent different route sets without fixed sequence information.

If no other road traffic route set is compatible with the urban rail-bound vehicle U_i, the compatible routes of the urban rail-bound vehicle U_i and road traffic route set R_{ROT,set_k} can keep the same operation at the level crossing, which means the road traffic route set R_{ROT,set_k} can extend its occupation time to the end of occupancy of the urban rail-bound vehicle U_i.

Otherwise, if some road traffic routes in route set $R_{ROT,set_{k+1}}$ compatible with the urban rail-bound vehicle U_i, based on the pre-defined sequential rule and separable principle (rules judgment with blue frame), it must be determined whether to allow the compatible road traffic routes in subset CR_{set_{k+1},U_i} to proceed to enter. In case the subset CR_{set_{k+1},U_i} is allowed to enter the level crossing, the temporal status is iterative to the state of the urban rail-bound vehicle U_i entering the level crossing in case 1.

b) Another situation is that the road traffic route set $R_{ROT,set_{RP}}$ exceeds its maximum red light phase when the urban rail-bound vehicle U_i proceeds to enter the investigated level crossing. Because of the inseparable principle, the ongoing road traffic route set R_{ROT,set_k} is exclusive to the urban rail-bound vehicle. The road traffic route set $R_{ROT,set_{RP}}$ has to wait until the road traffic route set R_{ROT,set_k} completes its minimum simulated effective occupation time $(min)SIMo_{ROT,set_k}$[41]. With the direct entry of the urban rail-bound vehicle U_i, it is possible that the time duration of the blocking time of the arriving urban rail-bound vehicle BL_{U_i} is shorter than the surplus time of the minimum simulated effective occupation time $(Sur)(min)SIMo_{ROT,set_k}$. The urban rail-bound vehicle U_i can leave the level crossing earlier without interrupting the road traffic.

Otherwise, the road traffic route set R_{ROT,set_k} completes its minimum simulated effective occupation time $(min)SIMo_{ROT,set_k}$ first. Then it is necessary to de-

---

[41] The minimum simulated effective occupation time here indicates two possibilities: 1) The road traffic route set has already completed its minimum green light phase. The green light has to be over right now. $(min)SIMo_{ROT,set_k}$ is the effective occupation time with the time duration of actual green light phase in simulation. 2) The minimum green light phase of the road traffic route set doesn't finish yet, which has to complete. $(min)SIMo_{ROT,set_k}$ is just the minimum effective occupation time.

termine whether the road traffic route set $R_{ROT,set_{RP}}$ exceeding its maximum red light phase is compatible with the urban rail-bound vehicle U_i. If they are compatible, the road traffic route set $R_{ROT,set_{RP}}$ is allowed to proceed to enter while the urban rail-bound vehicle U_i is running at the level crossing, which represents the state of this case a). If not, the road traffic route set $R_{ROT,set_{RP}}$ hast to wait until the urban rail-bound vehicle U_i leaves. The minimum simulated effective occupation time $(min)SIMo_{ROT,set_k}$ for the road traffic route set R_{ROT,set_k} can be extended till the urban rail-bound vehicle U_i leaves.

c) If the ongoing road traffic route set R_{ROT,set_k} is separable, the compatible route set $CR_{set_k,R_{set_{RP}}}$ can move with the road traffic route set $R_{ROT,set_{RP}}$ that exceeds its maximum red light phase. However, before the road traffic route set $R_{ROT,set_{RP}}$ can enter, the exclusive route set $ER_{set_k,R_{set_{RP}}}$ has to complete its $(min)SIMo_{ROT,set_k}$ first. If the urban rail-bound vehicle U_i leaves the level crossing in advance, the case is over and road traffic route set $R_{ROT,set_{RP}}$ proceeds to enter after the subset $ER_{set_k,R_{set_{RP}}}$ completes its $(min)SIMo_{ROT,set_k}$. The subset $ER_{set_k,R_{set_{RP}}}$ can complete its $(min)SIMo_{ROT,set_k}$ before the urban rail-bound vehicle U_i leaves the level crossing, it is possible that the urban rail-bound vehicle U_i is exclusive to the road traffic route set $R_{ROT,set_{RP}}$. Then the road traffic route set $R_{ROT,set_{RP}}$ cannot enter the level crossing until the urban rail-bound vehicle U_i leaves.

Otherwise, the road traffic route set $R_{ROT,set_{RP}}$ can enter with the set $CR_{set_k,R_{set_{RP}}}$. It is necessary to determine whether the time duration of the blocking time of the arriving urban rail-bound vehicle BL_{U_i} is shorter than the surplus effective occupation times of the other two road traffic route sets. If it is, the urban rail-bound vehicle can leave the level crossing first, the case is over. Otherwise, it has to determine whether there is other road traffic route set $CR_{set_{k+1},U_i}(\in R_{ROT,set_{k+1}})$ is compatible with the urban rail-bound vehicle U_i and the road traffic route sets. If no, it is able to extend the road traffic route set with shorter effective occupation time to the end of the blocking time of the arriving urban rail-bound vehicle BL_{U_i}. Otherwise, the road traffic route set

CR_{set_{k+1},U_i} is allowed to enter the investigated level crossing, which can be regarded as the initial state of this case c).

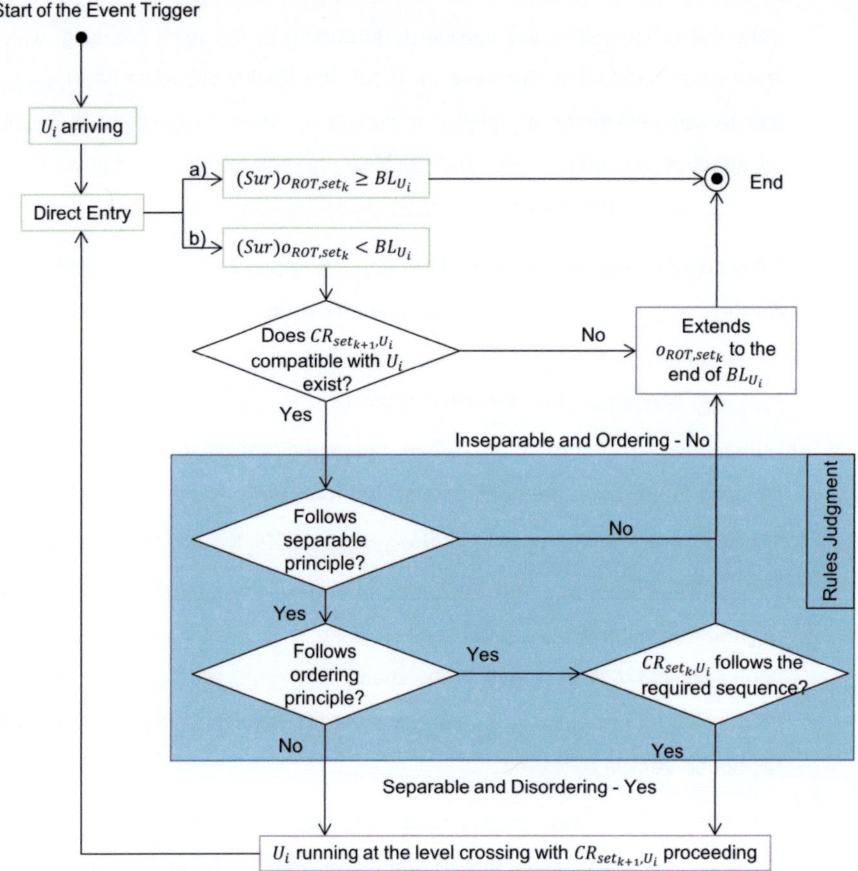

Figure 5-9: The Workflow of Execution of *LC* Case 1

LC Case 2: Road traffic routes in route set R_{ROT,set_k} is exclusive, and no other urban rail-bound transport is in progress

The ongoing road traffic route set R_{ROT,set_k} is exclusive to the urban rail-bound vehicle U_i when U_i running on its route $r_{URT,q}$ arrives at an investigated level crossing. Two statues for the road traffic route set R_{ROT,set_k} can be recognized with the event trigger condition of case 2. (see Appendix Figure 5)

a) The minimum green light phase $(min)GP_{ROT,set_k}$ of the road traffic route set R_{ROT,set_k} is not finished yet. The road traffic route set has to continue its operation until reaching the minimum green light phase and finishing its minimum effective occupation time $(min)o_{ROT,set_k}$ in simulation (also can be treated as $(min)SIMo_{ROT,set_k}$) before giving the way to the urban rail-bound vehicle U_i.

b) The road traffic route set R_{ROT,set_k} completes its minimum green light phase $(min)GP_{ROT,set_k}$ already. It has to proceeds to start its yellow light phase $YP(G)_{ROT,set_k}$ at once. This is the minimum simulated effective occupation time $(min)SIMo_{ROT,set_k}$[42].

The operation of road traffic route set R_{ROT,set_k} is interrupted and required to finish its minimum simulated effective occupation time $(min)SIMo_{ROT,set_k}$. Afterwards, the urban rail-bound vehicle U_i can have the right to proceed to enter the investigated level crossing while no other road traffic route set $R_{ROT,set_{RP}}$ waiting for the green light phase exceeds the limitation of maximum red light phase. Otherwise, whether the urban rail-bound vehicle U_i can gain the access to pass through the level crossing highly depends on the compatibility between the urban rail-bound vehicle U_i and the road traffic route set $R_{ROT,set_{RP}}$.

If the road traffic route set $R_{ROT,set_{RP}}$ exceeding the maximum red light phase is exclusive to the urban rail-bound vehicle U_i. The rail-bound vehicle has to wait while the road traffic route set $R_{ROT,set_{RP}}$ proceeds to enter the investigated level crossing, which is a similar initial state of this case. Otherwise, both of them can proceed to enter the investigated level crossing together. It is the state in the process of case 1.

However, it is also possible that the compatible road traffic route set CR_{set_k,U_i} can continue on the separable principle. The exclusive route set ER_{set_k,U_i} has to complete

---

[42] The simulated effective occupation time $SIMo_{ROT,set_k}$ here is the real-time effective occupation time of road traffic route set R_{ROT,set_k} in the simulation process of this iterative, which is calculated with the actual green light phase for R_{ROT,set_k} in simulation. It is called simulated effective occupation time $SIMo_{ROT,set_k}$. If the road traffic route have to finish its occupancy as soon as possible, the green light phase must finish immediately and then complete its other part of effective occupation time, which is the minimum simulated effective occupation time $(min)SIMo_{ROT,set_k}$.

its minimum simulated effective occupation time $(min)SIMo_{ROT,set_k}$ as well, and then the urban rail-bound vehicle U_i can gain the access to pass through the level crossing with the continuing subset CR_{set_k,U_i}, if there is no other road traffic route set $R_{ROT,set_{RP}}$ exceeding the limitation of maximum red light phase. The state with the urban rail-bound vehicle U_i and compatible road traffic route set CR_{set_k,U_i}, which is regarded as a process of case 1.

Otherwise, if the road traffic route set $R_{ROT,set_{RP}}$ exceeds the limitation of maximum red light phase, it has to determine whether the route set is compatible with the arriving urban rail-bound vehicle U_i. If it is compatible with both U_i and road traffic route set CR_{set_k,U_i}, it can proceed to enter together with the urban rail-bound vehicle U_i, and the compatible road traffic route set CR_{set_k,U_i} can also continue to complete its effective occupation time. If, it is only compatible with the urban rail-bound vehicle U_i, the road traffic route set R_{ROT,set_k} must complete its minimum simulated effective occupation time $(min)SIMo_{ROT,set_k}$ to give the right to the road traffic route set $R_{ROT,set_{RP}}$. Meanwhile, the urban rail-bound vehicle U_i gets access to enter the investigated level crossing too, which is further executed based on case 1. If the urban rail-bound vehicle U_i is exclusive to the road traffic route set $R_{ROT,set_{RP}}$ exceeding its limitation of maximum red light phase, the urban rail-bound vehicle has to wait as the initial state of this case. For the road traffic route set R_{ROT,set_k}, it is possible that a compatible subset $CR_{set_k,R_{set_{RP}}}$ exists and can continue with the road traffic route set $R_{ROT,set_{RP}}$ proceeding to enter.

LC Case 3: Road traffic route R_{ROT,set_k} and other urban rail-bound transport U_{i+1}[43] on its route $r_{URT,q+1}$ are compatible in progress

In this case, except the ongoing road traffic route set R_{ROT,set_k}, there is also other operating urban rail-bound vehicle U_{i+1} on its route $r_{URT,q+1}$ that is compatible with the arriving urban rail-bound vehicle U_i. The urban rail-bound vehicle U_i is allowed to

---

[43] There is no sequence meaning in timetable among the urban rail-bound transports U_i, U_{i+1}, ..., which means the possible urban rail-bound transports on its corresponding routes $r_{URT,q}$, $r_{URT,q+1}$, ... without sequence meaning of normal rotation in a traffic cycle.

proceed to enter the level crossing without extra waiting time and interruption on road traffic.

It is possible that there is the road traffic route set $R_{ROT,set_{RP}}$ exceeding its maximum red light phase at this moment. This route set may be exclusive to one of the ongoing road traffic route set R_{ROT,set_k} or urban rail-bound vehicle U_{i+1} or both of them. Moreover, the arriving urban rail-bound vehicle U_i may also be exclusive to or compatible with it. Therefore, in this case, there are seven possibilities (a) – g)) based on the actual simulation situation shown in Figure 5-10.

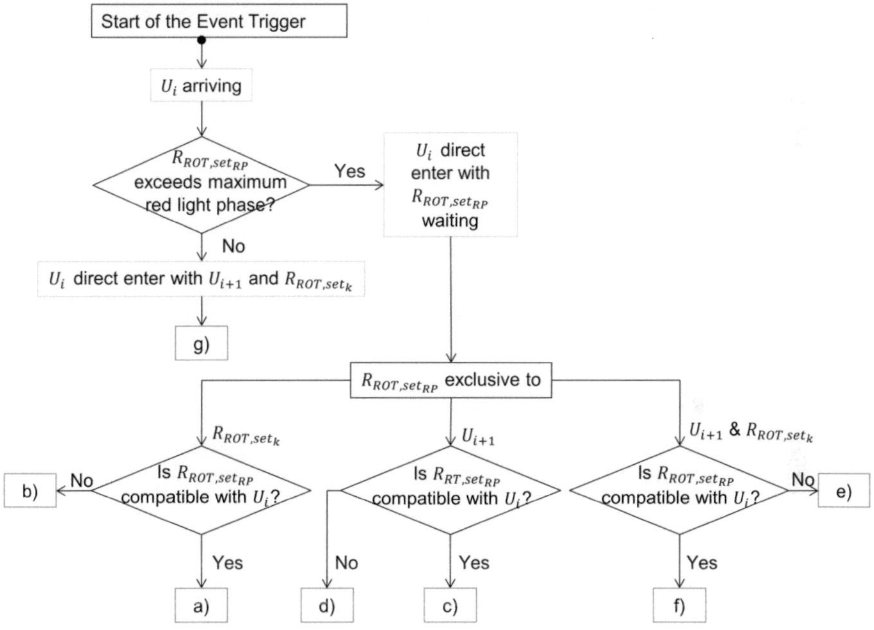

Figure 5-10: The Probabilities for LC Case 3

Besides, the probability that there is no road traffic route set $R_{ROT,set_{RP}}$ exceeding its maximum red light phase can show the basic logic of this case (see Appendix Figure 6.). Similar to that in case 1, the road traffic route set R_{ROT,set_k} can complete the surplus effective occupation time for road traffic route set $(Sur)o_{ROT,set_k}$, while the urban rail-bound vehicle U_{i+1} passes through the level crossing also needs surplus time of its blocking time of the level crossing $(Sur)BL_{U_{i+1}}$. There are three occurred scenarios when the urban rail-bound vehicle U_i passes through the level crossing during its entire blocking time BL_{U_i}. The other six possibilities in Figure 5-10 are extended based on the probability g), which determines the movement of the road traffic route set $R_{ROT,set_{RP}}$ exceeding its maximum red light phase respectively under the various three scenarios.

1) The surplus effective occupation time for road traffic route set $(Sur)o_{ROT,set_k}$ is longer than the (surplus) blocking time of both urban rail-bound vehicle BL_{U_i} and $(Sur)BL_{U_{i+1}}$. Under this situation, if the urban rail-bound vehicle U_i leaves the level crossing earlier than the urban rail-bound vehicle U_{i+1}, there will be no interference between urban rail-bound vehicle U_i and road traffic at the in-vestigated level crossing.

    The other situation is that the urban rail-bound vehicle U_{i+1} leaves the level crossing before the urban rail-bound vehicle U_i. There are one compatible road traffic route set R_{ROT,set_k} and urban rail-bound vehicle U_i running at the investigated level crossing. Similar to the judgments in case 1 and case 2, whether there is other road traffic route $r_{ROT,p+1}$ belonging to subset CR_{set_{k+1},U_i} compatible with both of them based on the relevant rules to determine the corresponding movements at the investigated level crossing.

2) The surplus effective occupation time for the road traffic route set $(Sur)o_{ROT,set_k}$ is shorter than the maximum (surplus) blocking time of two urban rail-bound vehicles BL_{U_i} and $(Sur)BL_{U_{i+1}}$, but longer than the minimum (surplus) blocking time of them. Similar to a) for this case, if the urban rail-bound vehicle U_i leaves earlier than the urban rail-bound vehicle U_{i+1}, there will be no interference between the arriving urban rail-bound vehicle U_i and road traffic at the investigated level crossing.

    Otherwise, it is possible that the blocking time of urban rail-bound vehicle BL_{U_i} is longer than the surplus blocking time of the urban rail-bound vehicle $(Sur)BL_{U_{i+1}}$. The urban rail-bound vehicle U_i and the road traffic route set R_{ROT,set_k} are still running at the investigated level crossing when the urban rail-bound vehicle U_{i+1} leaves. At this moment, it is possible that there is other road traffic route $r_{ROT,p+1}$ compatible with both of them. If no, the effective occupation time for the road traffic route o_{ROT,set_k} can be extended to the end of the urban rail-bound vehicle U_i, which can be regarded as the state in the process of case 1 with one urban rail-bound vehicle U_i and the road traffic route set R_{ROT,set_k} with a shorter surplus effective occupation time $(Sur)o_{ROT,set_k}$ running at the investigated level crossing.

If there is road traffic route $r_{ROT,p+1}$ compatible with both of them, it is necessary to determine which of the road traffic route set R_{ROT,set_k} and the road traffic route $r_{ROT,p+1}$ completes its occupancy first. It can determine the following movements of urban mixed traffic at the investigated level crossing.

3) The surplus effective occupation time for the road traffic route set $(Sur)o_{ROT,set_k}$ is completed first. By now, there are two urban rail-bound vehicles U_i and U_{i+1} at the investigated level crossing. Likewise, it is necessary to determine whether other road traffic routes compatible with the two urban rail-bound vehicle exist based on the relevant rules. If there is no other compatible road traffic route, the effective occupation time for the road traffic route o_{ROT,set_k} can be extended to the end of one of the urban rail-bound vehicles U_i and U_{i+1}. This situation can be treated as one of the states in the process of case 1 in which there are one urban rail-bound vehicle and one road traffic route set with a shorter surplus effective occupation time $(Sur)o_{ROT,set_k}$ running at the investigated level crossing.

Otherwise, it is an iterative process of the state of the urban rail-bound vehicle U_i entering the investigated level crossing in case 3 with the road traffic route $r_{ROT,p+1}$ and other urban rail-bound vehicle U_{i+1} running at the investigated level crossing compatible with it.

LC Case 4: Road traffic route set R_{ROT,set_k} is exclusive. Simultaneously, there is other urban rail-bound transport U_{i+1} on its route $r_{URT,q+1}$ in progress that is compatible

The event is triggered to execute the case 4 when there are road traffic route set R_{ROT,set_k} exclusive to while an urban rail-bound vehicle U_{i+1} compatible with the arriving urban rail-bound vehicle U_i. Appendix Figure 7 shows the interactions of urban mixed traffic in this case.

Similar to case 2, the urban rail-bound vehicle U_i is hindered by the ongoing road traffic route set R_{ROT,set_k}. It is necessary to determine whether the ongoing road traffic route set R_{ROT,set_k} completes its minimum green light phase to decide the corresponding surplus simulated effective occupation time of the road traffic route set $(Sur)o_{ROT,set_k}$.

a) The urban rail-bound vehicle U_{i+1} completes its surplus blocking time of $(Sur)BL_{U_{i+1}}$ earlier. When it leaves the investigated level crossing, the road traffic route set R_{ROT,set_k} still occupies the level crossing, and it is exclusive to the arriving urban rail-bound vehicle U_i. U_i still waits at the start point of the level crossing, which is precisely the status of the initial state of case 2.

b) The surplus effective occupation time of the road traffic route set R_{ROT,set_k} is over first. Similar to that in case 2, it is possible that there is the road traffic route set CR_{set_k,U_i} based on the separable principle that can continue to move with the arriving urban rail-bound vehicle U_i proceeding to enter if there is no other road traffic route set $R_{ROT,set_{RP}}$ exceeding its limitation of maximum red light phase.

Furthermore, based on the relevant rules, the determination whether there is other road traffic route set CR_{set_{k+2},U_i} compatible with the two urban rail-bound vehicles can be made. The compatible road traffic route set is allowed to proceed to enter with the urban rail-bound vehicle U_i. The status is treated as the state of the urban rail-bound vehicle U_i entering the level crossing in case 3.

Otherwise, the urban rail-bound vehicles U_i can proceed with only U_{i+1} running at the investigated level crossing. If the urban rail-bound vehicle U_{i+1} leaves early, it is necessary to determine whether other road traffic route set compatible with it exists. If there is none, the case is over with the urban rail-bound vehicle U_i leaving the level crossing and the road traffic continuing the normal rotation again. Otherwise, it is the temporal status that can be treated as the state of the urban rail-bound vehicle U_i entering the level crossing in case 1.

LC Case 5: Road traffic route set R_{ROT,set_k} is compatible. Simultaneously, there is other urban rail-bound transport U_{i+1} on its route $r_{URT,q+1}$ in progress that is exclusive

This case is with the opposite of the initial state in case 4. There is other urban rail-bound vehicle U_{i+1} exclusive to the arriving urban rail-bound vehicle U_i. However, the ongoing road traffic route set R_{RT,set_k} is compatible with this urban rail-bound vehicle U_i. Because of the exclusive urban rail-bound vehicle U_{i+1}, U_i must stop and wait at the start point of the investigated level crossing until the exclusive urban rail-bound vehicle U_{i+1} leaves the level crossing. The logic is shown in Figure 5-11.

a) If the urban rail-bound vehicle U_{i+1} leaves the level crossing in advance, The road traffic route set R_{ROT,set_k} is compatible with the urban rail-bound vehicle U_i. It becomes the temporal status similar to the urban rail-bound vehicle U_i entering the level crossing in case 1.

b) The surplus effective occupation time of the road traffic route set $(Sur)o_{ROT,set_k}$ is shorter, which means it finishes before the urban rail-bound vehicle U_{i+1} leaves the level crossing. If there is other road traffic route set CR_{set_{k+1},U_i+1} compatible with the urban rail-bound vehicle U_{i+1} according to the sequential and route set rules. The compatible route set is allowed to proceed to start its green light phase with the ongoing urban rail-bound vehicle U_{i+1}. Depending on the compatibility between the road traffic route set CR_{set_{k+1},U_i+1} and the urban rail-bound vehicle U_i, the following reactions for the urban rail-bound vehicle U_i can be can be taken after the urban rail-bound vehicle U_{i+1} leaves. If they are compatible with each other, the situation becomes the temporal status of the initial state in case 5 iteratively. Otherwise, it is the temporal status iterative to the initial state of case 6.

If there is no other road traffic route compatible with the urban rail-bound vehicle U_{i+1}, the road traffic route set R_{ROT,set_k} can extend its occupancy until the end of the blocking time of the urban rail-bound vehicle U_{i+1}. At this moment, the situation becomes the initial state of case 1 with an arriving urban rail-bound vehicle and ongoing road traffic route set compatible with each other.

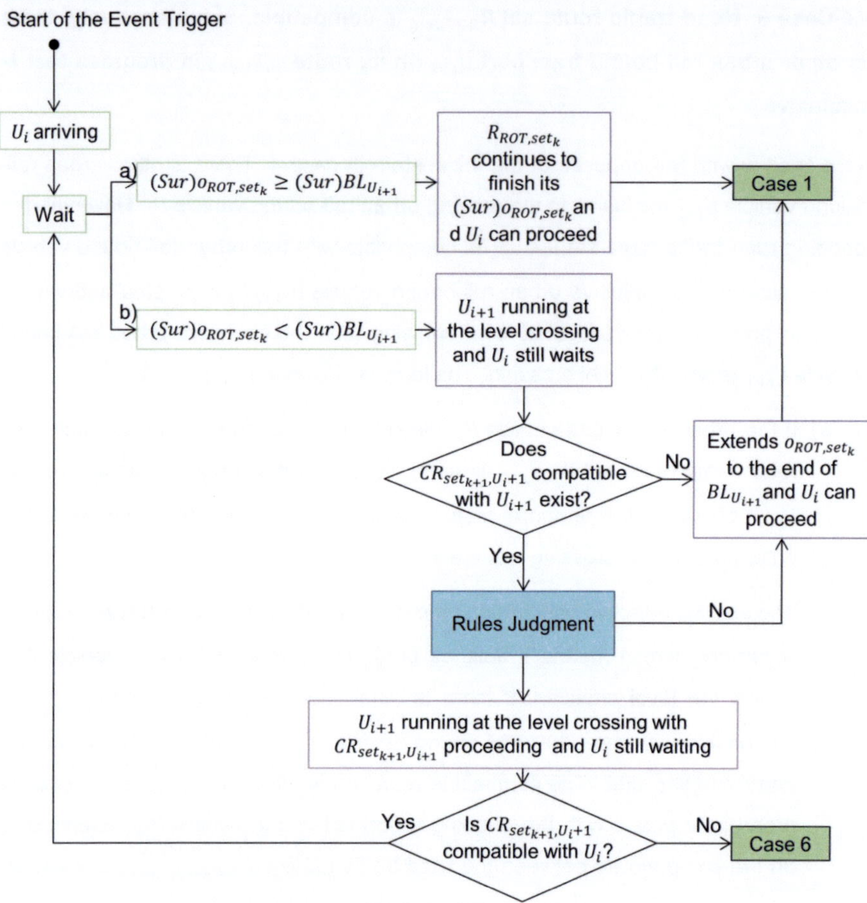

Figure 5-11: The Workflow of Execution of *LC* Case 5

LC Case 6: Road traffic route set R_{ROT,set_k} and other urban rail-bound transport U_{i+1} on its route $r_{URT,q+1}$ in progress are exclusive

The event to execute case 6 is triggered, when an urban rail-bound vehicle U_i on its route $r_{URT,q}$ arrives at the investigated level crossing. Both the ongoing road traffic route set R_{ROT,set_k} and other urban rail-bound vehicle U_{i+1} are exclusive to the arriving urban rail-bound vehicle U_i. As described in cases 2 and 4, the ongoing road traffic route set R_{ROT,set_k} must complete its minimum (simulated) effective occupation time as well. The reactions are differently with the duration of the surplus minimum (simulated) effective occupation time of road traffic route set $(Sur)o_{ROT,set_k}$ and the surplus blocking time of urban rail-bound vehicle $(Sur)BL_{U_{i+1}}$ as shown in Figure 5-12.

a) The surplus minimum (simulated) effective occupation time of road traffic route set $(min)SIMo_{ROT,set_k}$ is longer than the surplus blocking time of urban rail-bound vehicle $(Sur)BL_{U_{i+1}}$. The road traffic route set R_{ROT,set_k} continues to complete its surplus minimum (simulated) effective occupation time of road traffic route set $(Sur)SIMo_{ROT,set_k}$ and the arriving urban rail-bound vehicle U_i needs to wait at the start point, which is the initial state of case 2.

b) The road traffic route set R_{ROT,set_k} completes its minimum (simulated) effective occupation time of road traffic route set $(min)SIMo_{ROT,set_k}$ in advance. It can extend its minimum (simulated) effective occupation time due to the urban rail-bound vehicle U_{i+1} that is also exclusive to the arriving urban rail-bound vehicle U_i.

If the surplus blocking time of urban rail-bound vehicle $(Sur)BL_{U_{i+1}}$ finishes before the road traffic route set R_{ROT,set_k}, the exclusive urban rail-bound vehicle U_{i+1} leaves the investigated level crossing first. Therefore, the situation can be treated as the initial state of case 2 with an ongoing road traffic route set and an urban rail-bound vehicle U_i waiting at the start point.

Otherwise, the effective occupation time of road traffic route set o_{ROT,set_k} can be extended to the end of the blocking time of urban rail-bound vehicle $(Sur)BL_{U_{i+1}}$. At this moment, if there is no other road traffic route set exceeding its maximum red light phase, the urban rail-bound vehicle U_i can proceed to enter the investigated level crossing. Otherwise, the outcome is determined by the compatibility of the road traffic route set $R_{ROT,set_{RP}}$ exceeding its maximum red light phase and the urban rail-bound vehicle U_i. If they are exclusive to each other, the urban rail-bound vehicle U_i has to wait with the road traffic route set entering, which can be treated as the initial state of case 2.

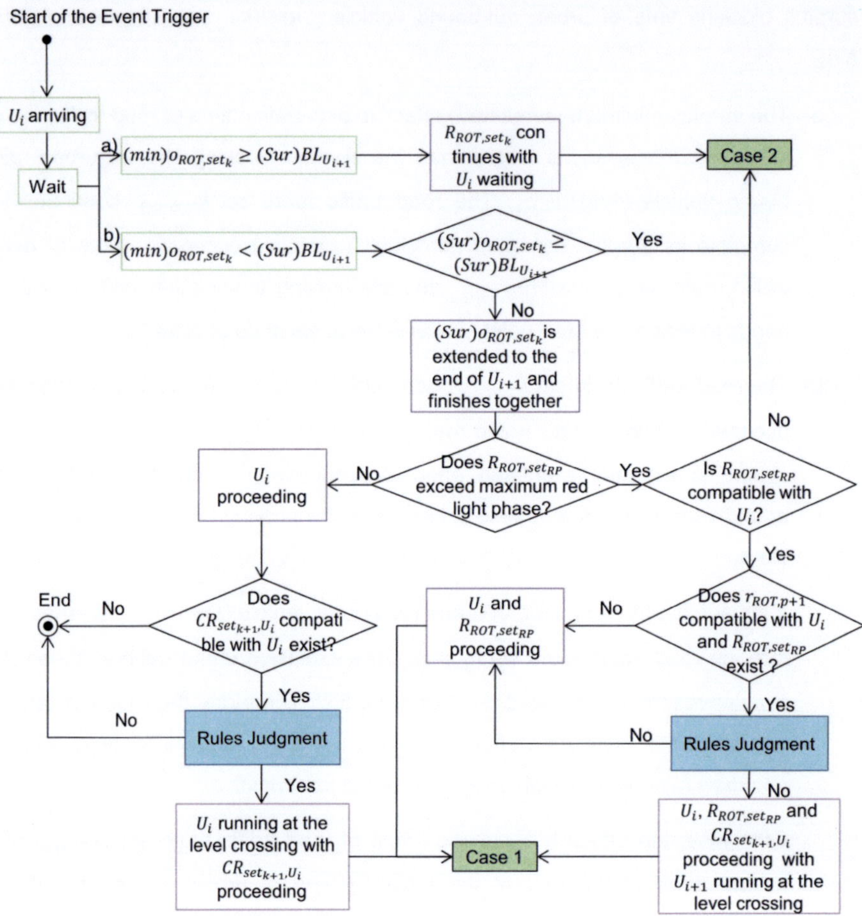

Figure 5-12: Workflow of Execution of *LC* Case 6

5.3.3   *SR* Cases 1 and 2 for Shared Road

Along the mixed traffic zone of shared road, the urban rail-bound vehicle is restricted once it is hindered by road traffic at the investigated shared road at a random time point (t) during the investigated time period (T), which means the event is triggered.

As described in Subchapter 5.2, the urban rail-bound vehicle U_i with the semi-random entry time $T_{URT,i}$: $(00{:}00{:}T_{URT,i})$ is hindered by road traffic ROT_j at a random location point $LP(t)$: $(s_{ROT,j}, t_{ROT,j})$ with a running distance of urban rail-bound vehicle $s_{URT,i}$: $(x_{URT,i}, 0)$ and a corresponding running time $t_{URT,i}$: $(00{:}00{:}t_{URT,i} + T_{URT,i})$. The running distance of urban rail-bound vehicle $s_{URT,i}$ is just the location information of the encountered road traffic $s_{ROT,j}$, while the time course t from the entry time point of the hindered urban rail-bound vehicle is equal to the corresponding time information of the road traffic $t_{ROT,j}$ and the running time of urban rail-bound vehicle $t_{URT,i}$ as shown in Figure 5-12.

There are two possible responses to the urban rail-bound vehicle U_i.

- The urban rail-bound vehicle U_i follows the road traffic ROT_j at its corresponding speed $V_{ROT,j}$ without congestion in front
- The urban rail-bound vehicle U_i follows the road traffic ROT_j at its corresponding speed $V_{ROT,j}$ with congestion in front

Whether road traffic ROT_j is congested or not depends on a given possible proportion on all random road traffic generated along the investigated shared road. Meanwhile, the corresponding speed of road traffic ROT_j is also randomly determined by the given proportion of various road traffic types at their speeds. The road traffic ROT_j is generated randomly with the information of its location point $LP(t): (s_{ROT,j}, t_{ROT,j})$ along the shared road in each second during the investigated time period.

The urban rail-bound vehicle U_i can be hindered by different road traffic with different speeds in each iterative for Θ times. When the road traffic hinders the urban rail-bound transport with a lower speed at the θ th iterative, the urban rail-bound transport has to follow the lower speed. The urban rail-bound transport is hindered by congested road traffic before the next level crossing or the simulation is terminated at the end of the investigated time period. Once the hindrance by congestion occurs, the urban rail-bound transport follows the road traffic controlled by the corresponding traffic control signal at the next level crossing. Consequently, the corresponding hindrance time of urban rail-bound vehicle U_i caused by road traffic along the investigated shared road can be calculated (see Subchapter 5.4).

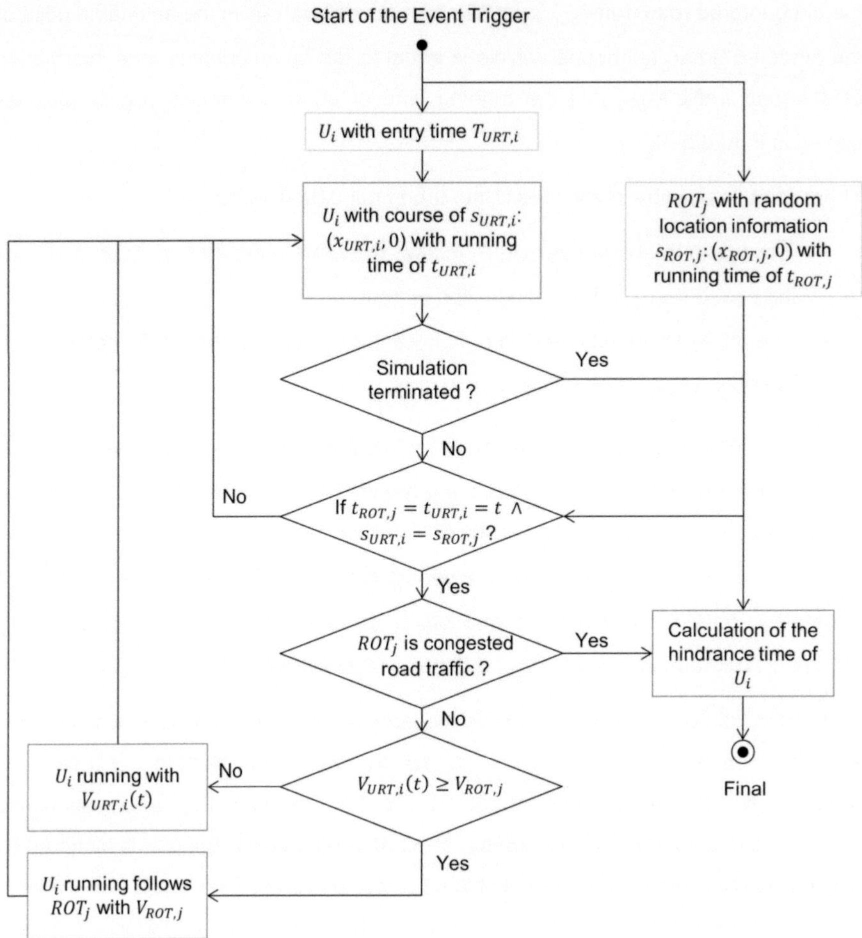

Figure 5-13: Workflow of *SR* Case for Mixed Traffic Zone of Shared Road

If there is stop on the shared road, in this algorithm, the investigated shared road is divided as two connected mixed traffic zones of shared road by the stop. Because road traffic can influence the urban rail-bound vehicle U_i even at the area of the stop position, the urban rail-bound vehicle U_i exits from the first shared road when it reaches the defined location point for the platform. After the dwell time, it is the situation that the urban rail-bound vehicle U_i proceeds to enter the next shared road with. These two connected shared roads are indirectly controlled by the same traffic control signal at the next level crossing.

If the congested road traffic are in the green light phase, the hindered urban rail-bound vehicle U_i can follow the road traffic to run until this green light phase is over. It has to wait for $(TC - GP)$[44] until the next green light phase of the traffic control signals at the next adjacent level crossing proceeds again and lasts for the time duration of green light phase GP. This is an iterative process until the urban rail-bound vehicle U_i exits the shared road. The green light phase starts randomly in the first traffic cycle with starting time T_{GP} from the entry time point of the hindered urban rail-bound transport during the investigated time period.

The urban rail-bound transport can only run one by one along the investigated shared road if there is more than one urban rail-bound transport at the same time. The influences between urban rail-bound transports can be reflected by the semi-randomly created entry time point for each urban rail-bound transport when it enters the shared road. If there is other urban rail-bound vehicle U_{i+1} also running along the shared road following the urban rail-bound vehicle U_i, the urban rail-bound vehicle U_i in front is treated as road traffic in the real simulated time, and it may hinder the urban rail-bound vehicle U_{i+1}.

5.3.4 SS Cases 1 and 2 for Shared Space

The case for mixed traffic zone of shared space is different from that of shared road. The influences of pedestrians' movements or stops of cars and buses on urban rail-bound vehicle U_i are iterated and superimposed for Θ iterative with assumed average waiting time $WT_{URT,i}$ in each hindrance.

The urban rail-bound vehicle U_i with its semi-random departure time $T_{URT,i}$: $(00{:}00; T_{URT,i})$ created based on time slice can be hindered by road traffic ROT_j at a random location point $LP(t)$: $(s_{ROT,j}, t_{ROT,j})$. For the θ th iterative, the running distance of urban rail-bound vehicle $s_{URT,i}^{\theta}$ contains location coordinate ($x_{URT,i}^{\theta} = \sum_{\theta=1}^{\theta} s_{URT,i}^{\theta}, 0$) and corresponding running time $t_{URT,i}^{\theta}$ at the time point $(00{:}00{:} \sum_{\theta=1}^{\theta}(t_{URT,i}^{\theta} + WT_{URT,i}^{\theta-1}) + T_{URT,i})$. The pedestrians' movements and the tem-

---

[44] $TC - GP$: The difference between the time duration of all non-green light phases in one traffic cycle TC and the corresponding green light phase (GP) is the time period that, the urban rail-bound transport following the road traffic cannot run on the investigated shared road.

porary stops of cars can be modeled as random road traffic. In addition, the scheduled stops of buses are modeled as some fixed points with time and location point information. The urban rail-bound vehicle U_i must stop for the assumed average waiting time $WT_{URT,i}$ before it restarts again. It is considered that in each iterative θ (i.e. hindrance) the running distance of urban rail-bound vehicle $s_{URT,i}^{\theta}$ is updated based on the pre-defined operation of it, in order to better represent the dynamics of urban rail-bound transport. Therefore, the simulated running distance (coordinate for the urban rail-bound vehicle U_i) can be calculated by adding up each running distance in Θ iterative. The simulated running time is also the sum of each running time and the extra waiting time as well.

As shown in Figure 5-14, through judging whether the urban rail-bound vehicle U_i ($s_{URT,i}^{\theta}, t_{URT,i}^{\theta}$) is hindered by the road traffic ROT_j ($s_{ROT,j}, t_{ROT,j}$), the corresponding responses can be determined. The simulation won't be terminated until the investigated time period completes. On the shared space, similar to the shared road, the urban rail-bound vehicle can only run consecutively even though there are lots of trains simultaneously. The semi-random created entry time point for each urban rail-bound transport can show the interferences between various urban rail-bound transports entering the shared space. Once there is other urban rail-bound vehicle U_{i+1} running along the shared space behind the urban rail-bound vehicle U_i, the urban rail-bound vehicle U_i is considered as a road traffic that hinders the urban rail-bound vehicle U_{i+1} with the surplus waiting time at the simulated time point.

Therefore, based on the recorded results from simulation, it is able to derive the hindrance time of urban rail-bound vehicle U_i caused by road traffic along the investigated shared space (see Subchapter 5.4).

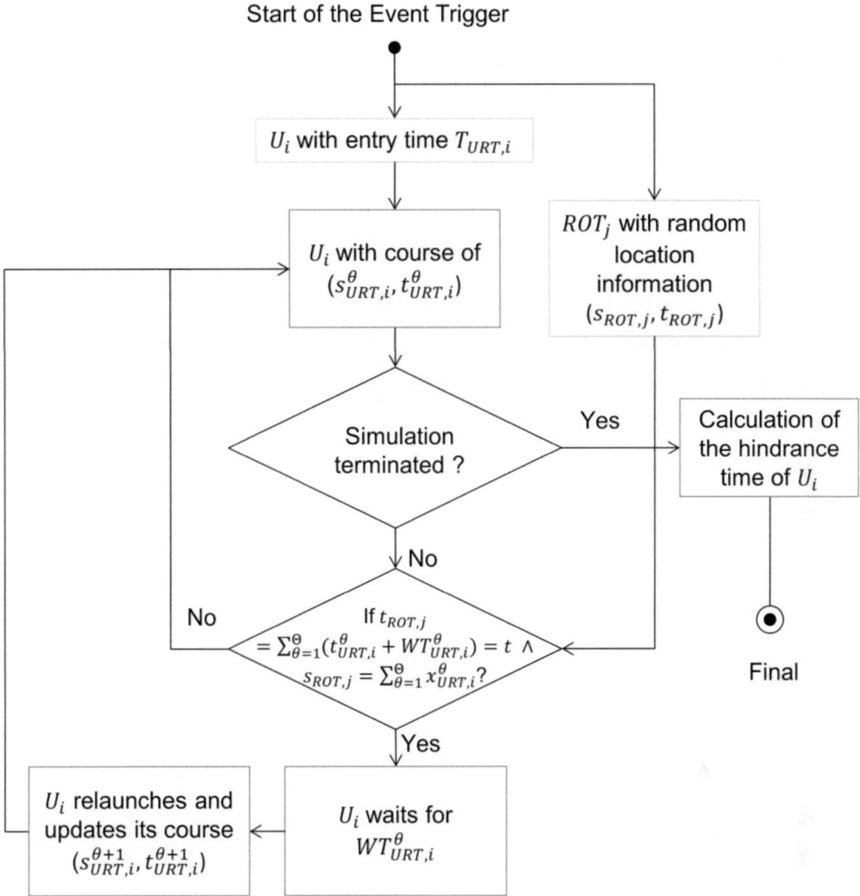

Figure 5-14: Workflow of SS Case for Mixed Traffic Zone of Shared Space

5.4 Calculation of Indicators

In order to determine the road traffic influences at various mixed traffic zones, the corresponding indictors have to be determined using the simulation results based on the algorithm with event-driven system described in Subchapter 5.3. For various types of mixed traffic zones, in this dissertation the indicators are explained and calculated in this subchapter respectively.

5.4.1 Indicators of Algorithm for Level Crossing

In order to derive the results of capacity research with the consideration of road traffic influences at mixed traffic zone of level crossing, two indictors have to be determined, as described in the DFG-research project [Martin & Liu 2016a]:

- Potential occupation time of road traffic (O_{ROT})
- Hindrance time of urban rail-bound transport caused by road traffic (H_{URT})

Moreover, based on the reactions of the corresponding cases described in Subchapter 5.3.2, the indicator for the arriving urban rail-bound vehicle U_i on its route $r_{URT,q}$ can be determined by adding up the blocking time (BL_{U_i}) and the simulated hindrance time caused by both of road traffic and other urban rail-bound transports at the investigated level crossing.

- The total running simulated time of urban rail-bound vehicle U_i along the investigated level crossing $T_{SIM,i}$

The occupation time of road traffic set for a whole traffic cycle can be obtained based on the simulation results for each simulated effective occupation time of road traffic route set $SIMo_{ROT,set_k}$ in a traffic cycle that is the actual effective occupation time of road traffic route set in simulation of the developed modeling with event-driven system. Therefore, the sum of these simulated effective occupation time during an investigated time period at an investigated level crossing can be calculated as the indicator of potential occupation time of road traffic (O_{ROT}). If the separable principle is considered, the simulated effective occupation time of road traffic route set $SIMo_{ROT,set_k}$ is calculated as the simulated effective occupation time of its subsets with the longest occupation.

The value, as described in [Martin & Liu 2016a], is considered as the maximal possible occupation time of road traffic on the corresponding route for an investigated.

$$O_{ROT} = \sum_{c=1}^{C} \sum_{k=1}^{K} SIMo_{ROT,set_k} \qquad (\,5\text{-}10\,)$$

With:

$SIMo_{ROT,set_k}$ The simulated effective occupation time of road traffic route set R_{ROT,set_k} that is exclusive to others in one traffic cycle at an investigated mixed traffic zone

k	The index of the route set of road traffic routes with $k \in [1, K]$
K	The number of route sets in a traffic cycle
C	The number of traffic cycles in an investigated time period T
c	The index of the traffic cycles with $c \in [1, C]$

It can be seen that the road traffic occupy the level crossing all the time (except the all red light phase) if there is no urban rail-bound transport at the investigated level crossing. The value of the potential occupation time for road traffic (O_{ROT}) should be equal to the time duration of the investigated time period (T) when there is no all red light phase theoretically [Martin & Liu 2016a].

Therefore, if one or more urban rail-bound transports are scheduled to pass through the level crossing, the extra influences caused by road traffic will increase, leading to the hindrance time of urban rail-bound transports. Meanwhile, the potential occupation time for road traffic (O_{ROT}) may go down. The hindrance time of an urban rail-bound transport on its route $r_{URT,q}$ by a road traffic route set R_{ROT,set_k} in a traffic cycle can be determined based on the simulation results from the model of event-driven system in this algorithm. If the road traffic route set R_{ROT,set_k} is separable, the hindrance time of the urban rail-bound vehicle U_i on its route $r_{URT,q}$ (h_{URT,i,set_k} that is h_{URT,q,set_k}) depends on the hindrance time by those separated subsets of the road traffic route set R_{ROT,set_k} in a traffic cycle.

During the occupancy of the road traffic route set R_{ROT,set_k}, one or more urban rail-bound transports might be hindered by the route set or its subsets (separable principle). Therefore, the effective hindrance time (h_{URT,set_k}) of urban rail-bound transports caused by the road traffic route set R_{ROT,set_k} or its subsets in one traffic cycle can be derived by the maximum hindrance time of all urban rail-bound transports on theirs routes (based on [Martin & Liu 2016a]).

$$h_{URT,set_k} = max \left\{ h_{U_i,q,set_k}, h_{U_j,q+1,set_k}, \cdots \right\} \qquad (5\text{-}11)$$

With:

h_{URT,set_k}	The effective hindrance time of urban rail-bound transports on theirs routes caused by road traffic route set R_{ROT,set_k} during one traffic cycle at an investigated level crossing

h_{U_i,q,set_k} / $h_{U_j,q+1,set_k}$ — The hindrance time of urban rail-bound vehicle U_i on its route $r_{URT,q}$ or U_j on its route $r_{URT,q+1}$ caused by road traffic route set R_{ROT,set_k}, with $i, j, …$ the index of urban rail-bound transport on its corresponding route

q — The index of the urban rail-bound transport routes, $q \in [1, n]$

set_k — The index of the road traffic route sets, $set_k \in [1, m_{set_k}]$

In [Martin & Liu 2016a], the indicator of the hindrance time of urban rail-bound transport caused by road traffic (H_{URT}) at an investigated level crossing in an investigated time period was defined as:

"The total occupation time of road traffic, at which the urban rail-bound transports are hindered by the occupation of road traffic at the investigated mixed traffic zone during an investigated time period."

The value (H_{URT}) can be calculated from the effective hindrance time (h_{URT,set_k}) of urban rail-bound transports on the route caused by road traffic set R_{ROT,set_k} in each traffic cycle during the investigated time period.

$$H_{URT} = \sum_{c=1}^{C} \sum_{set_k=1}^{K} h_{URT,set_k} \qquad\qquad (\,5\text{-}12\,)$$

With:

k — The index of the route set of road traffic routes with $k \in [1, K]$

K — The number of road traffic route sets in a traffic cycle

C — The number of traffic cycles in an investigated time period T

c — The index of the traffic cycles with $c \in [1, C]$

Based on all cases described in Subchapter 5.3, the interactions between road traffic and urban rail-bound transport(s) at a mixed traffic zone of level crossing can be modeled in an established event-driven system with the relevant rules and inputs introduced in Subchapter 5.2. Consequently, the above interpreted indicators can be derived from the simulation results through the developed algorithm. Furthermore, the results of capacity research, waiting time function and throughput capacity can be determined based on the indicators (see Subchapter 5.5).

5.4.2 Indicators of Algorithm for Shared Road

For the mixed traffic zone of shared road, as description in Subchapter 5.3.3, the urban rail-bound transport is hindered by low speed road traffic or by congested road traffic as well. Therefore, there are two possibilities for the urban rail-bound transport:

- Hindered only by road traffic with lower speed
- Hindered by road traffic with congestion as well

In order to derive the results for capacity research on shared road, the hindrance (i.e. waiting time) of urban rail-bound transport can be calculated with the difference between the simulated transport time determined by the simulated exit time point and the scheduled entry time point (00:00:00) and the scheduled transport time [Schmidt 2009]:

$$WT_{URT,i} = \left(t_{SIM,exit} - t_{Schedule,entry}\right) - T_{B,i} \qquad (5\text{-}13)$$

With:

$WT_{URT,i}$ — The waiting time of urban rail-bound vehicle U_i along the investigated shared road

$T_{B,i}$ — The scheduled transport time of the urban rail-bound vehicle U_i along the shared road

$t_{SIM,exit}$ — The simulated exit time point of the urban rail-bound vehicle U_i

$t_{Schedule,entry}$ — The scheduled entry time point of the urban rail-bound vehicle U_i

Therefore, it is necessary to determine the simulated exit time point using the model of this algorithm in event-driven system. For the situation with hindrance only from the lower speed road traffic, the indicator can determine the simulated exit time point $t_{SIM,exit}$ with the simulated semi-random entry time $T_{URT,i}$: $(00\!:\!00\!:\!T_{URT,i})$ has to be determined:

- The total simulated running time of urban rail-bound vehicle U_i along the investigated shared road $T_{SIM,i}$

Moreover, for the situation with the hindrance caused by the congested road traffic, another indicator for the determination of the indicator of the total running time of urban rail-bound vehicle U_i along the investigated shared road $T_{SIM,i}$ is:

- The hindrance time from congested road traffic $(Cong)H_{URT,i}$

If the urban rail-bound transport is hindered only by the random road traffic with lower speed in front, it has to follow the movement of the road traffic with the lowest speed. Based on the model described in Subchapter 5.3.3, the road traffic ROT_j with the information of its location $LP(t)$: $(s_{ROT,j}, t_{ROT,j})$ is generated randomly along the shared road during the investigated time period T. There are Θ iterative in which the urban rail-bound vehicle U_i encounters randomly generated road traffic with corresponding information $(s_{ROT,j}^{\theta}, t_{ROT,j}^{\theta})$[45].

The urban rail-bound transport will follow the lowest speed $\min_{\theta \in \Theta}(V_{ROT,j}^{\theta})$ of all these encountered road traffic. Therefore, the indicator of the total simulated running time $T_{SIM,i}$ of urban rail-bound vehicle U_i along the investigated shared road can be determined based on the length of the investigated shared road:

$$T_{SIM,i} = t_{ROT,j}^{\min(V_{RT,j}^{\theta})} + \frac{S - s_{ROT,j}^{\min(V_{ROT,j}^{\theta})}}{\min\limits_{\theta \in \Theta}(V_{ROT,j}^{\theta})} \tag{5-14}$$

With:

$T_{SIM,i}$	The total simulated running time of urban rail-bound vehicle U_i along the investigated shared road
$t_{ROT,j}^{\min(V_{ROT,j}^{\theta})}$	The time (point) of the encountered road traffic hindering the urban rail-bound vehicle U_i, which moves at the lowest speed
$s_{ROT,j}^{\min(V_{ROT,j}^{\theta})}$	The (moving) distance of the encountered road traffic hindering the urban rail-bound vehicle U_i, which moves at the lowest speed
S	The length of the investigated shared road

---

[45] It can indicate the real simulated time point and running distance with corresponding coordinate for an urban rail-bound transport ahead of the urban rail-bound transport U_i, which is hindered by other road traffic and accordingly hinders the urban rail-bound transport U_i with lower speed.

$$\min_{\theta \in \Theta} (V_{ROT,j}^{\theta})$$ The minimum speed of the encountered road traffic hindering the urban rail-bound vehicle U_i

Accordingly, the simulated exit time point $t_{SIM,exit}$ can be further derived by adding the random entry time point of the urban rail-bound vehicle $T_{URT,i}$: $(00{:}00{:}T_{URT,i})$ with the calculated total simulated running time $T_{SIM,i}$ of urban rail-bound vehicle U_i along the investigated shared road. Therefore, the extra waiting time caused by road traffic along the shared road can be determined for the situation with the encountered road traffic at lower speed.

In other situations, the urban rail-bound vehicle U_i is also hindered by congested road traffic. The influences of road traffic can be calculated with the hindrance time of urban rail-bound transports at the hindered point by the congested road traffic. The urban rail-bound transport is controlled by the traffic control signal at the next adjacent level crossing while it passes the rest part of the investigated shared road at the lowest speed of the encountered road traffic.

If the urban rail-bound transport encounters road traffic congestion, the traffic light phase will be crucial for the movements of the urban rail-bound transports at the lowest speed of the encountered road traffic. Therefore, the surplus green light phase should be applied when the urban rail-bound vehicle U_i encounters the congested road traffic with $((Cong)s_{ROT,j}^{\theta}, (Cong)t_{ROT,j}^{\theta})$.

$$(Sur)GP_{URT,i} = \left(\left(\left\lfloor \left| \frac{(Cong)t_{ROT,j}^{\theta}}{TC} \right| \right\rfloor \right) \times TC + T_{GP} + GP \right) - (Cong)t_{ROT,j}^{\theta} \quad (\,5\text{-}15\,)$$

With:

$(Sur)GP_{URT,i}$ — The surplus green light phase for urban rail-bound vehicle U_i when it is hindered by the congested road traffic along the investigated shared road

$(Cong)t_{ROT,j}^{\theta}$ — The time (point) of the congested road traffic hindering the urban rail-bound vehicle U_i

TC — The time duration of a traffic cycle of traffic control signal at the next adjacent level crossing

T_{GP} — The time (point) of the green light phase in the first traffic cycle of traffic control signal at the next level crossing during

GP an investigated time period

GP The time duration of the corresponding green light phase in a traffic cycle of the traffic control signal for the corresponding connected route at the next level crossing

This calculated surplus green light phase can be positive or negative, which is determined by whether the time point of hindrance $(Cong)t_{ROT,j}^{\theta}$[46] is at the green light phase or other traffic light phases.

If the surplus green light phase is larger than zero ($(Sur)GP_{URT,i} > 0$), the time point of hindrance $(Cong)t_{ROT,j}^{\theta}$ is still at the green light phase. Once this green light phase completes, the urban rail-bound transport has to stop and wait for the next green light phase of traffic control signal of the corresponding connected route at the next level crossing. Therefore, the number of hindrances that the urban rail-bound vehicle U_i after the surplus green light phase has to stop and wait for the next green light phase can be calculated as:

$$h = \left\lfloor \frac{S - \left((Cong)s_{ROT,j}^{\theta} + \min_{\theta \in \Theta} (V_{ROT,j}^{\theta}) \cdot (Sur)GP_{URT,i}\right)}{GP \cdot \min_{\theta \in \Theta} (V_{ROT,j}^{\theta})} \right\rfloor + 1 \qquad (\,5\text{-}16\,)$$

With:

h The number of hindrances that the urban rail-bound transport experiences after encountering the congested road traffic with index of ε

S The length of the investigated shared road

$(Cong)s_{ROT,j}^{\theta}$ The (moving) distance of the congested road traffic hindering the urban rail-bound vehicle U_i

$\min_{\theta \in \Theta} (V_{ROT,j}^{\theta})$ The minimum speed of the encountered road traffic hindering the urban rail-bound vehicle U_i

Therefore, the indicator of the hindrance time of urban rail-bound vehicle U_i $(Cong)H_{URT,i}$ is caused by congested road traffic at the investigated shared road when the encountered time point is at the green light phase (i.e. $(Sur)GP_{URT,i} > 0$).

---

[46] This is the time point of the urban rail-bound transport hindered by the congested road traffic, which is also the temporal time course $t = (Cong)t_{ROT,j}^{\theta} = t_{URT,i}$ from its entry time point.

This hindrance time can be calculated as the sum of the time duration of all hindrances during which the urban rail-bound vehicle U_i stops and waits for the next green light phase after the surplus green light phase.

$$(Cong)H_{URT,i} = \sum_{\varepsilon=1}^{h}(TC - GP)$$

(5-17)

With:

$(Cong)H_{URT,i}$ The indicator of the hindrance time of urban rail-bound vehicle U_i by congested road traffic at the investigated shared road

TC The time duration of a traffic cycle of traffic control signal at the following adjacent level crossing

GP The time duration of the corresponding green light phase in a traffic cycle of the traffic control signal of the corresponding connected route at the next level crossing

On the other hand, if the surplus green light phase is less than zero ($(Sur)GP_{URT,i} \leq 0$), the time point when the urban rail-bound vehicle U_i is hindered by the congested road traffic $(Cong)t^{\theta}_{ROT,j}$ is just at other traffic light phases rather than green light phase. This means that the urban rail-bound vehicle U_i has to stop and wait for the next green light phase. Therefore, it is necessary to determine the surplus waiting time $(Sur)(TC - GP)^{(Cong)t^{\theta}_{ROT,j}}_{URT,i}$ of it for the next green light phase, which may be less than or equal to all other traffic light phases except the corresponding green light phase $(TC - GP)$.

$$(Sur)(TC - GP)^{(Cong)t^{\theta}_{ROT,j}}_{URT,i} = TC - GP - \left|(Sur)GP_{URT,i}\right|$$

(5-18)

With:

$(Sur)(TC - GP)^{(Cong)t^{\theta}_{ROT,j}}_{URT,i}$ The surplus waiting time of the urban rail-bound vehicle U_i hindered by the congested road traffic for the next green light phase

$(Sur)GP_{URT,i}$[47]   The surplus green light phase for urban rail-

---

[47] In this situation, it is less than zero.

bound vehicle U_i when it is hindered by the congested road traffic along the investigated shared road

Similarly, if the urban rail-bound vehicle U_i is hindered by the congested road traffic, the encountered time point is at other traffic light phases rather than the corresponding green light phase (i.e. $(Sur)GP_{URT,i} \leq 0$). The number of hindrances can be calculated using the following formula.

$$h = \left| \frac{S - (Cong)s_{ROT,j}^{\theta}}{GP \cdot \min_{\theta \in \Theta} (V_{ROT,j}^{\theta})} \right| + 1 \qquad (\,5\text{-}19\,)$$

With:

h	The number of hindrances that the urban rail-bound transport experiences after encountering the congested road traffic with index of ε
S	The length of the investigated shared road,
$(Cong)s_{ROT,j}^{\theta}$	The (moving) distance of the congested road traffic hindering the urban rail-bound vehicle U_i
$\min_{\theta \in \Theta} (V_{ROT,j}^{\theta})$	The minimum speed of the encountered road traffic hindering the urban rail-bound vehicle U_i
GP	The time duration of the corresponding green light phase in a traffic cycle of the traffic control signal of the corresponding connected route at the next level crossing

Consequently, the indicator of the hindrance time of urban rail-bound vehicle U_i by congested road traffic $(Cong)H_{URT,i}$ at the investigated shared road can be calculated as:

$$(Cong)H_{URT,i} = (Sur)(TC - GP)_{URT,i}^{(Cong)t_{ROT,j}^{\theta}} + \sum_{\varepsilon=1}^{h-1}(TC - GP) \qquad (\,5\text{-}20\,)$$

With:

$(Cong)H_{URT,i}$	The indicator of the hindrance time of urban rail-bound vehicle U_i by congested road traffic at the investigated shared road
TC	The time duration of a traffic cycle of traffic control signal at

the next adjacent level crossing

In conclusion, the urban rail-bound vehicle U_i is influenced not only by the road traffic with lower, but also by the congestion of road traffic in front controlled by the traffic control signal of the corresponding connected route at the next adjacent level crossing. Therefore, the indicator of the total simulated running time of urban rail-bound vehicle U_i along the investigated shared road $T_{SIM,i}$ in this situation shoule be determined using the following formula.

$$T_{SIM,i} = t_{ROT,j}^{\min(V_{ROT,j}^\theta)} + \frac{S - s_{ROT,j}^{\min(V_{ROT,j}^\theta)}}{\min\limits_{\theta \in \Theta}(V_{ROT,j}^\theta)} + (Cong)H_{URT,i} \qquad (\,5\text{-}21\,)$$

With:

$T_{SIM,i}$ The total simulated running time of urban rail-bound vehicle U_i along the investigated shared road

$t_{RT,j}^{\min(V_{ROT,j}^\theta)}$ The time (point) of the encountered road traffic hindering the urban rail-bound vehicle U_i, which moves at the lowest speed

$s_{RT,j}^{\min(V_{ROT,j}^\theta)}$ The (moving) distance of the encountered road traffic hindering the urban rail-bound vehicle U_i, which moves at the lowest speed

S The length of the investigated shared road

$\min\limits_{\theta \in \Theta}(V_{ROT,j}^\theta)$ The minimum speed of the encountered road traffic hindering the urban rail-bound vehicle U_i

$(Cong)H_{URT,i}$ The indicator of the hindrance time of urban rail-bound vehicle U_i by congested road traffic at the investigated shared road

As shown above, it is able to calculate the simulated exit time point $t_{SIM,exit}$, based on the randomly created entry time point of the urban rail-bound vehicle $T_{URT,i}$: $(00{:}00{:}T_{URT,i})$ and the determined total simulated running time $T_{SIM,i}$ of urban rail-bound vehicle U_i along the investigated shared road. Accordingly, the road traffic influences on the urban rail-bound vehicle U_i can be used for further capacity research of evaluation.

5.4.3   Indicators of Algorithm for Shared Space

For the mixed traffic zone of shared space, the model is implemented in an iterative process with the possible hindrance times as Θ. For each iterative, the urban rail-bound vehicle U_i needs to restart. Therefore, the corresponding running time and distance are determined based on the pre-defined dynamics of urban rail-bound transports. Therefore, it is able to determine the indicator for deriving the extra waiting time of urban rail-bound transport caused by road traffic on the investigated shared space, which can further determine the simulated exit time point $t_{SIM,exit}$ through adding to the simulated semi-random entry time $T_{URT,i}: (00:00:T_{URT,i})$ as described in Subchapter 5.4.2:

- The total simulated running time of urban rail-bound vehicle U_i along the investigated shared space $T_{SIM,i}$

After the urban rail-bound vehicle U_i is hindered by road traffic for the last time, it will be the $\theta = \Theta$ th iterative. There may also be some surplus part of shared space $(Sur)S$ where urban rail-bound transport runs U_i without road traffic influences.

$$(Sur)S = S - x_{URT,i}^{\theta=\Theta} = S - \sum_{\theta=1}^{\Theta} s_{URT,i}^{\theta} \qquad (\,5\text{-}22\,)$$

With:

$(Sur)S$ The surplus length of the investigated shared space after the urban rail-bound vehicle U_i hindered by all road traffic along the shared space

Θ The number of the hindrances (iterative)

θ The index of the number of the hindrances with $\theta \in [1, \Theta]$

$x_{URT,i}^{\theta=\Theta}$ The coordinate of urban rail-bound vehicle U_i when it is hindered by road traffic for $\theta = \Theta$ iterative

S The length of the investigated shared space

$s_{URT,i}^{\theta}$ The running distance of urban rail-bound vehicle U_i encounters the θ th hindrance from the location point of the $(\theta - 1)$ th hindrance

Based on the pre-defined dynamics of urban rail-bound transport, the corresponding running time for the urban rail-bound vehicle U_i to pass through the surplus length of the investigated shared space $(Sur)S$ is known as $(Sur)t_{URT,i}$. Meanwhile, the running time of urban rail-bound vehicle U_i encounters the θ th hindrance from the time point of the $(\theta - 1)$ th hindrance is known as $t_{URT,i}^{\theta}$. Therefore, the indicator of the total simulated running time $T_{SIM,i}$ of urban rail-bound vehicle U_i along the investigated shared space can be determined with the assumed waiting time of urban rail-bound vehicle $WT_{URT,i}^{\theta}$[48] for the θ th hindrance.

$$T_{SIM,i} = (Sur)t_{URT,i} + \sum_{\theta=1}^{\Theta} \left(t_{URT,i}^{\theta} + WT_{URT,i}^{\theta} \right) \qquad (\ 5\text{-}23\ )$$

With:

$T_{SIM,i}$ The total simulated running time of urban rail-bound vehicle U_i along the investigated shared space

$(Sur)S$ The surplus length of the investigated shared space after the urban rail-bound vehicle U_i hindered by all road traffic along the shared space

$t_{URT,i}^{\theta}$ The running time of urban rail-bound vehicle U_i when it is the θ th hindrance caused by road traffic

$WT_{URT,i}^{\theta}$ The assumed waiting time of urban rail-bound vehicle U_i for the θ th hindrance of road traffic

Similarly, according to the randomly created entry time point of the urban rail-bound vehicle $T_{URT,i}$: $(00{:}00{:}T_{URT,i})$, the simulated exit time point $t_{SIM,exit}$ for urban rail-bound vehicle U_i to exit the investigated shared space can be determined by adding it up with the calculated total simulated running time $T_{SIM,i}$ of urban rail-bound vehicle U_i along the investigated shared space.

In summary, with the developed algorithm for the model of automatic event-driven system, the extra waiting time caused by road traffic influences of all urban rail-bound transports running in the investigated mixed traffic zone during the investigated time period based on the give operating program (scheduled timetable) can be determined. Moreover, a large number of random cases can be generated for determina-

---

[48] The waiting time can also be the hindrance time caused by the previous urban rail-bound transport that is hindered by random road traffic or other urban rail-bound transports. The time duration of the waiting time is the surplus waiting time of the previous hindered urban rail-bound transport.

tion of the indicators for capacity research. The average of these random results from simulation shows the possible road traffic influences at the investigated mixed traffic zone in reality, which is plausible with higher explanatory power to a certain extent.

During the whole investigated time period, the number of urban rail-bound transport is determined based on the operating program (scheduled timetable). The simulations with this model of event-driven system are executed for the whole investigated time period. Therefore, the average waiting time of urban rail-bound transports caused by road traffic ($(ROT)ETw$) can be calculated for mixed traffic zone of shared road and shared space with respective total simulated running time and the scheduled transport time of urban rail-bound vehicle U_i along the investigated mixed traffic zone:

$$(ROT)ETw = \frac{\sum_{i=1}^{N}(T_{SIM,i} - T_{B,i})}{N}$$

(5-24)

With:

$(ROT)ETw$ The extra waiting time of urban rail-bound vehicle U_i caused by road traffic along the investigated mixed traffic zone of shared road or shared space

$T_{SIM,i}$ The total simulated running time of urban rail-bound vehicle U_i along the investigated mixed traffic zone of shared road or shared space

$T_{B,i}$ The scheduled transport time of the urban rail-bound vehicle U_i along the investigated mixed traffic zone of shared road or shared space

For mixed traffic zone of level crossing, the hindrance time at the investigated level crossing is caused by both road traffic and other urban rail-bound transports. Therefore, the waiting time of urban rail-bound vehicle U_i caused by road traffic can be determined with the difference of the total simulated running time of urban rail-bound vehicle U_i with road traffic influences ($T_{SIM,i}$ described in Subchapter 5.4.1) and the total simulated running time of urban rail-bound vehicle U_i without road traffic influences[49] under the same situation (the same random entry time point of urban rail-bound vehicle U_i based on the given operating program). Therefore, it is able to derive the average waiting time of all urban rail-bound transport $((ROT)ETw)$ caused by road traffic at the investigated level crossing.

5.5 Determination of Results for Capacity Research

Based on simulation results from the model with event-driven system of the developed algorithm, the average waiting time of urban rail-bound transports with the influences of road traffic at an investigated mixed traffic zone can be determined as described in Subchapter 5.4. Accordingly, the average waiting time of urban rail-bound transports with stepwise-varied traffic flows based on the given operating program can also be derived for the determination of waiting time function. For the throughput capacity, the entry traffic flow and the exit traffic flow during the evaluated time period can also be derived for urban rail-bound transports with stepwise-varied traffic flows from the simulation results.

Therefore, with regard to an investigated mixed traffic zone as the investigated area, the corresponding results (waiting time function and throughput capacity) for capacity research to evaluate the operating performance can be determined and used to further derive the recommended area of traffic flow. The results can also reflect the significance of the road traffic influences at the investigated mixed traffic zone. It is very useful for the assessment of the impact of the road traffic on urban rail-bound transports in the whole investigated area with multiple mixed traffic zones. It is more valuable to study the road traffic influences in reality. Even though there are generally more than one mixed traffic zones in an urban rail-bound network, some mixed traffic zones have minor impact on urban rail-bound transports. In consideration of the efforts of cost and time, such mixed traffic zones can be negligible for capacity research of the whole investigated area. Therefore, it is important to carry out the preliminary assessment with the developed algorithm, which can balance the accuracy of results and the cost of evaluation.

---

[49] The total simulated running time of urban rail-bound transport U_i without road traffic influences can be obtained from the simulation results with the help of simulation tool RailSys and the assistant software PULEIV.

The DFG-research project [Martin & Liu 2016a] proposed the method to determine the waiting time function approximatively based on the results of capacity research without taking into consideration of road traffic influences. Regardless of whether it's an investigated mixed traffic zone or a whole investigated area, the determined three parameters (a_2, b_2, c_2) for the waiting time function (4-7) without road traffic influences are used as the parameters (a_5, b_5, c_5) for the waiting time function (4-11) in a simple way. Because the waiting time function determines the fitting curve of the discrete points, it is obvious that the parameters will be changed. There will be errors on the results of waiting time function (4-11). However, according to the request, the approximate result can represent the road traffic influences on urban rail-bound transport to a certain extent.

Therefore, there is only one parameter that have to be derived for the waiting time function with road traffic influences for the investigated mixed traffic zone, with the determined throughput capacity in advance [Martin & Liu 2016a]:

- $d\ (= d_5/(1-\eta)^{2/3})$: The starting point (waiting time) of waiting time function (4-11) with road traffic influences.

In the following two subchapters, determination of capacity research results for a single investigated mixed traffic zone and the whole investigated area are discussed respectively.

5.5.1 Examining a Single Mixed Traffic Zone

When the road traffic influences at the investigated mixed traffic zone are taken into account, the throughput capacity $(\widehat{DS\,LF})$ will be reduced due to the hindrances of road traffic comparing with that without road traffic influences. The parameter d $(= d_5/(1-\eta)^{2/3})$ in the waiting time function (4-11) represents the average waiting time caused by road traffic influences if there is only one urban rail-bound transport without hindrances between other urban rail-bound transports at the investigated mixed traffic zone.

In order to derive relatively accurate results for capacity research, it is able to use the average waiting time of urban rail-bound transports with stepwise-varied traffic flows at the investigated mixed traffic zone determined by simulated results (see Subchapter 5.4). The waiting time function (4-11) can be fitted with these data points and the throughput capacity with road traffic influences ($\widehat{DS\,LF}$) at the investigated mixed traffic zone. The throughput capacity is determined with the algorithm proposed by [Schmidt 2009] and developed by [Chu 2014] using the numbers of entering and exiting urban rail-bound transport with stepwise-varied traffic flows that can be obtained from the simulation results of the event-driven system modeling with the developed algorithm. The results derived in this way do have with higher accuracy.

On the other hand, according to the requirement, it is also possible to derive the parameter d $(= d_5/(1 - \eta)^{2/3})$ to define the waiting time function (4-11) with road traffic influences using the determined parameters (a_2, b_2, c_2) in the waiting time function (4-7) in the case without road traffic influences, as described in [Martin & Liu 2016a].

For an investigated mixed traffic zone of level crossing, the relationship of the two calculated indicators with the modeling of event-driven system, the potential occupation time for road traffic (O_{ROT}) and the hindrance time of urban rail-bound transport caused by road traffic (H_{URT}) shown in Figure 5-15, can derive the throughput capacity. According to the pre-defined relevant rules and the interactions between urban rail-bound transports and road traffic modeling in the event-driven system, the potential occupation time for road traffic (O_{ROT}) increases along with the increased stepwise-varied traffic flows of urban rail-bound transport that compete for occupation of mixed traffic zone of level crossing. Due to the competitive relation between urban rail-bound transport and road traffic at the investigated level crossing during an investigated time period, with more urban rail-bound transports entering the level crossing, the urban rail-bound transports are hindered much more by road traffic with reduced occupancy.

It is understandable that the potential occupation time for road traffic (O_{ROT}) can be as low as the minimum value, which means the road traffic at the investigated level crossing can occupy the level crossing with the minimum effective occupation time $(min)o_{ROT,p}$ in the stationary phase of the evaluated time period theoretically when the traffic flow of urban rail-bound transport is high. In practice, it is hard to reach this theoretical minimum value, because road traffic on the route may be compatible, the actual occupancy of the road traffic route is not absolutely equal to but mild longer than the minimum effective occupation time $(min)o_{ROT,p}$. Under this situation, urban rail-bound transports with much higher traffic flow cannot pass through the level crossing during the investigated time period, which is restricted by the traffic control signal rules of limitations of maximum red light phase for road traffic. The urban rail-bound transport can increase to the maximum traffic flow that is the approximate throughput capacity with road traffic influences $(\widehat{DS\,LF})$ [Martin & Liu 2016b].

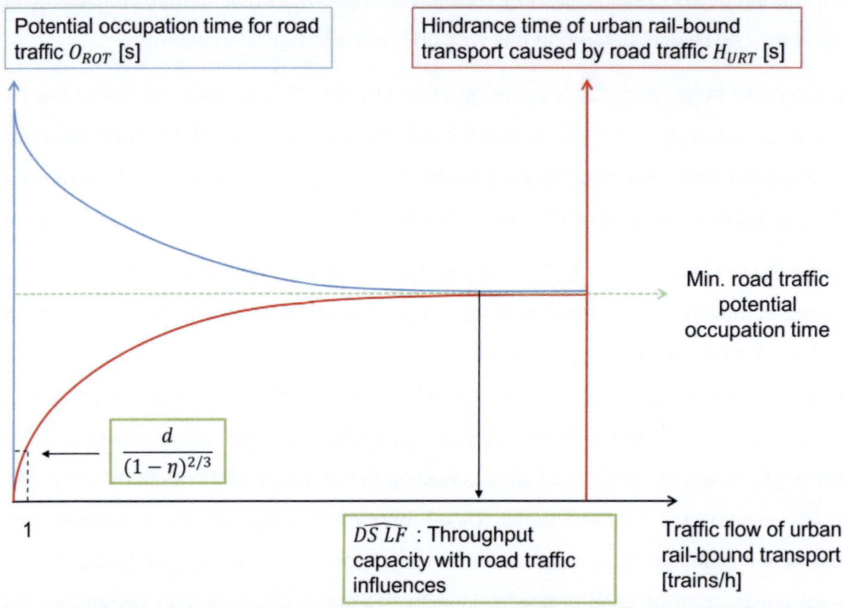

Figure 5-15: Relationship between Potential Occupation Time for Road Traffic (O_{ROT}) and the Hindrance Time of Urban Rail-bound Transport by Road Traffic (H_{URT}) (Source: [Martin & Liu 2016a] modified based on [Martin & Liu 2016b])

The parameter d $(= d_5/(1-\eta)^{2/3})$ for waiting time function (4-11) with road traffic influences is the starting point (waiting time) of the waiting time function, which indicates the hindrance time of the urban rail-bound transport with the traffic flow in the range of (0, 1] caused by road traffic at the investigated level crossing. The value of the hindrance time for the only one urban rail-bound transport U_i caused by road traffic (H_{URT}) can illustrate the probability of the occurring waiting time due to the operational hindrance by the influences of road traffic at the level crossing [Martin & Liu 2016b].

This concept is also suitable for other types of mixed traffic zones of shared road and shared space. The hindrance time (i.e. waiting time) of urban rail-bound transports caused by road traffic with one urban rail-bound transport was introduced in Subchapters 5.4.2 and 5.4.3. Due to the interdependent rather than competition along the shared road and the random iterative hindrances along the shared space, the throughput capacity with road traffic influences $(\widehat{DS\,LF})$ can be calculated based on the algorithm through comparison of the number of trains entering and exiting the investigated mixed traffic zone during the evaluated time period with the increase of stepwise-varied traffic flows of urban rail-bound transport based on the operating program [Chu 2014].

Throughput capacity and further derived recommended area of traffic flow for a single mixed traffic zone with road traffic influences (example in Appendix I) using the described two approaches in Chapter 3 and the developed algorithm are shown in Table 5-2. These capacity research results have an obvious lower comparing with those without road traffic influences in the same railway infrastructure and the given operating program for urban rail-bound transports.

Scenarios	Without Road Traffic Influences	With Road Traffic Influences		
		Modeling Approach	Distribution Approach	Event-driven System
Throughput Capacity	137.5 trains/h	80.5 trains/h	84.3 trains/h	82 trains/h
Recommended Area of Traffic Flow	74.2 – 104.1 trains/h	46.4 – 58.7 trains/h	48.6 – 60.5 trains/h	46.3 – 59.6 trains/h

Table 5-2: Capacity Research Results of a Single Mixed Traffic Zone Using Various Methods[50]

With the event-driven system, a large amount of samples are generated to be simulated, and the average of simulation results can represent the probability of the random influences of road traffic on urban rail-bound transport in an investigated mixed traffic zone during the evaluated time period. The average value of the indicators for mixed traffic zone (Subchapter 5.4) can be calculated and recorded to derive the results of capacity research. As previously mentioned, if necessary, the waiting time of each urban rail-bound transport in the model of this algorithm can also be gathered statistics, and then with the derived throughput capacity $(\widehat{DS\,LF})$ to fit the curve of waiting time function (4-11) with road traffic influences. Consequently, the signifi-

---

[50]It shows the result of an example of mixed traffic zone of level crossing that introduced in Appendix I: Basic Information of the Investigated Example.

cance of the road traffic influences in the investigated mixed traffic zone can be assessed.

5.5.2   Examining the Whole Investigated Area

Through the preliminary study on one specific mixed traffic zone, the assessment is executed. According to the demand, the selected significant mixed traffic zones in which the road traffic influences are relatively significant show that it is important to determine the results of capacity research for the whole investigated area with road traffic influences in these selected mixed traffic zones.

For a whole investigated area with one or more selected mixed traffic zones, the capacity research can be carried out integrating with the distribution approach (see Subchapter 3.3). For the distribution approach, the first step of capacity research is to determine the perturbation of the road traffic influences. As mentioned in Subchapter 5.4, the waiting time of urban rail-bound transports caused by road traffic in the investigated mixed traffic zone can be derived, which can be regarded as the perturbation caused by road traffic influences in the corresponding mixed traffic zone and is added into the simulation model of the whole investigated area. Therefore, it is possible to complete capacity research with distribution approach, to determine the throughput capacity ($\widehat{DS\,LF}$), the waiting time function (4-11) with road traffic influences, and eventually derive the recommended area of traffic flow for the whole investigated area.

On the other hand, it is also possible to conduct capacity research for the whole investigated area based on [Martin & Liu 2016a]. This concept is utilized for a specific mixed traffic zone as described in Subchapter 5.5.1. The waiting time function with road traffic influences at those selected mixed traffic zones can be approximately determined based on the parameters (a_2, b_2, c_2) in the waiting time function (4-7) resulted by capacity research without road traffic influences for the whole investigated area.

The throughput capacity with road traffic influences ($\widehat{DS\,LF}(MTZ_i)$) for each selected mixed traffic zone (MTZ_i) can be determined with the developed algorithm as described in Subchapter 5.5.1 as well as the parameter ($d(MTZ_i)$) of the starting point of waiting time function (4-11) for mixed traffic zone (MTZ_i). The throughput capacity is defined as the maximum capacity with unchanged operating program, while the mixed traffic zones with road traffic influences on urban rail-bound transport can be treated as possible bottlenecks in the whole investigated area, which limits the maximum capacity of the whole investigated area. As discussed in the DFG-research project [Martin & Liu 2016a], the scheduled traffic flow $(STF(MTZ_i))$ of urban rail-bound transport at each selected mixed traffic zone in a whole investigated area should be compared with the corresponding throughput capacity with road traffic influences $(\widehat{DS\,LF}(MTZ_i))$, the mixed traffic zone with the lowest increase of traffic flow of urban rail-bound transport is considered as the most affected node that is the bottleneck for the whole investigated area.

$$\widehat{DS\,LF}(W) = \min_{mtz}\left\{\frac{\widehat{DS\,LF}(MTZ_i)}{STF(MTZ_i)}\right\} \times STF \qquad (\,5\text{-}25\,)$$

With:

$\widehat{DS\,LF}(MTZ_i)$ The throughput capacity with road traffic influences for the mixed traffic zone MTZ_i in the whole investigated area W

$STF(MTZ_i)$ The scheduled traffic flow of urban rail-bound transport in the mixed traffic zone MTZ_i in the whole investigated area W defined in the scheduled timetable based on the given operating program

STF The traffic flow of all urban rail-bound transports in the scheduled timetable

MTZ_i One of the selected mixed traffic zones in the whole investigated area W

i The index of the mixed traffic zones in the whole investigated area W with the number of mixed traffic zones of mtz, $i \in [1, mtz]$

Furthermore, it is reasonable that the parameter $(d(MTZ_i))$ of waiting time function (4-11) for each mixed traffic zone (MTZ_i) in the whole investigated area can be used to derive the parameter $(d(W) = d_5/(1 - \eta)^{2/3})$ of the waiting time function (4-11) for the whole investigated area with the corresponding pre-defined recovery time $(REC(MTZ_i))$ of each mixed traffic zone (MTZ_i) in scheduled timetable [Martin & Liu 2016a].

$$d(W) = \sum_{i=1}^{mtz} (d(MTZ_i) - REC(MTZ_i)) \qquad\qquad (5\text{-}26)$$

With:

$d(MTZ_i)$ The parameter of waiting time function (4-11) for the mixed traffic zone MTZ_i in the whole investigated area W

mtz The number of the mixed traffic zones in the whole investigated area W, $i \in [1, mtz]$

$REC(MTZ_i)$ The pre-defined recovery time of urban rail-bound transport for the mixed traffic zone MTZ_i

Consequently, the adapted waiting time function with road traffic influences for the whole investigated area can be approximately determined with the model of this algorithm in an event-driven system. Accordingly, the recommended area of traffic flow can be derived for evaluation of operating performance.

Therefore, similar to the application to the case of a single mixed traffic zone, there are some errors unavoidable, the results of this algorithm for the whole investigated area can approximately present the influences caused by road traffic in the whole investigated area. According to the demand, the results from integrating with the distribution approach, which was described previously for a whole investigated area, can be obtained with higher accuracy. As shown in Table 5-3, the results of capacity research illustrate that the limitation of road traffic influences on the possible maximum capacity of urban rail-bound transport is also considerable in the whole investigated area with mixed traffic zones. The recommended area of traffic flow with the consideration of road traffic influences also reduces comparing to that in the pure rail-bound system without external influences.

Scenarios	Without Road Traffic Influences	With Road Traffic Influences		
		Modeling Approach	Distribution Approach	Event-driven System
Throughput Capacity	146.3 trains/h	104.3 trains/h	103.1 trains/h	102.5 trains/h
Recommended Area of Traffic Flow	93.9 – 139.8 trains/h	60.3 – 91.7 trains/h	63.6 – 92.2 trains/h	61.6 – 93.5 trains/h

Table 5-3: Capacity Research Results of a Whole Investigated Area Using Various Methods[51]

5.6   Conclusion

In this chapter, an algorithm for capacity research with road traffic influences in an urban rail-bound network is developed based on the DFG-research project [Martin & Liu 2016a]. It can help cut the cost and time on data collection. The behaviors of road traffic can be reflected relatively more exactly comparing with the two approaches described in previous Chapter 3.

This algorithm can be implemented only with the basic information of the investigated area. The model of event-driven system can be built for a specific mixed traffic zone. The results of capacity research can be derived directly using the calculated indicators that are determined by the simulation results under various cases when an event is triggered in the model. According to the requirements, the waiting time function can be also determined based on the partly known parameters determined by the waiting time function without road traffic influences at the investigated area with relatively lower accuracy but less efforts.

Furthermore, the developed algorithm in this chapter is a useful assessment method to determine the significances of the road traffic influences in the mixed traffic zones on a whole investigated area preliminarily. The cost and time can be reduced with the algorithm as pre-assessment to exclude the low influenced mixed traffic zones in a large-scale investigated area for capacity research of urban rail-bound transport with consideration of road traffic influences using other developed approaches [Martin & Liu 2016a].

---

[51] It shows the result of an example of mixed traffic zone of level crossing that introduced in Appendix I: Basic Information of the Investigated Example.

6 Summary and Future Development

In this dissertation, various urban mixed traffic zones were studied for the capacity research of urban rail-bound transport with consideration of external influences caused by road traffic. Based on the findings of [Hertel 1992] and further developments with simulation method of capacity research in [Schmidt 2009] and [Chu 2014], this research is an expansion of the existing capacity research for railway system.

- The urban rail-bound transport is influenced by road traffic in an investigated urban mixed traffic zone, leading to an increase of waiting time and an amount of capacity loss of urban rail-bound transport. The basic categories and characteristics of urban mixed traffic and urban mixed traffic zones are identified preliminarily. It is valid for the investigation of the capacity research on the urban rail-bound transport with road traffic influences.

- The mutual interactions among urban mixed traffic result in the external influences on the operation of urban rail-bound transport, and those interactions are thoroughly analyzed for description of the road traffic influences on urban rail-bound transport under various conditions. It is interpreted that the multifarious interactions in various investigated mixed traffic zones under specific condition of control and coordination of traffic control signals.

- On the basis of the analysis of urban mixed traffic, two approaches were developed to study the road traffic influences on urban rail-bound transport for capacity research. The stochastic influences caused by road traffic can be described systematically in the simulation model of urban rail-bound transport with proper approach according to the required application conditions and acquired data.

  For the modeling approach, the road traffic influences are directly reflected in the simulation of the trains operation of "modeled road traffic" and "trains" of urban rail-bound transports in the investigated time period. It is suitable for a relative simple and small-scale investigated area, especially with the urban mixed traffic zone of level crossing under control of traffic control signals.

  On the other hand, the distribution approach is appropriate for the complex and large-scale investigated area with various types of mixed traffic zones. It is not restricted with the structure of the investigated area as long as the required statistic databases of delays are sufficient for the determination of the probability distribution for perturbations representing road traffic influences. The determined perturbation is imported into the simulation model on the operation of urban rail-bound transport to derive the road traffic influences.

- For the urban mixed traffic pattern, the waiting time function is the key intermediate result of capacity research. The model function fits the data points (average waiting time) from the simulations of the stochastic influenced timetables with stepwise-varied traffic flows. Due to the stochastic influences of road traffic, the adapted model function of waiting time function (4-11) is developed based on the existing function form, which can fit the discrete data points much better (with higher (adjusted) coefficient of determination) and accordingly can derive the recommended area of traffic flow with higher explanatory power.

- As a prerequisite, it is important to determine whether an urban mixed traffic zone in a whole investigated area is worthy and necessary to be analyzed with the two developed approaches. In order to balance the cost, time and the required accuracy of results, the algorithm is developed to assess all mixed traffic zones in the whole investigated area in advance and then some mixed traffic zones with higher impacts of road traffic influences can be selected for the further study of capacity research.

  In addition, the developed algorithm for the modeling of an event-driven system with the basic defined rules and given inputs can also directly determine the results of capacity research – throughput capacity and waiting time function based on the simulation results under some assumptions. It can be applied for a specific investigated mixed traffic zone or the whole investigated area with one or more mixed traffic zones.

The research on urban rail-bound transport in this dissertation is relevant new. The existing findings are rarely related to the urban mixed traffic of urban rail-bound transport and road traffic. Therefore, some points can be further studied in future developments.

The capacity research on urban rail-bound transport is based on the existing simulation method. Accordingly, the sufficient simulations for determination of the waiting time function are necessary, especially, four parameters in the newly adapted waiting time function (4-11) have to be determined. From mathematical view, it will be difficult to be fitted with higher degree of freedom. As a matter of fact, it is considered that the adapted waiting time function (4-11) can be further investigated with another parameter in the newly added term $d_5/(1-\eta)^{e_5}$ to reflect the change of the road traffic influences with the increase of traffic flow of urban rail-bound transport more accurately in the future research.

In addition, for the road traffic, there are some assumptions on the random and complex behaviors of road traffic that were simplified in this research. Such as for the mixed traffic zone of shared space, the road traffic with various trip purposes and behaviors may respond differently when an urban rail-bound transport is approaching. This can be further analyzed in details. Moreover, for the various special conditions mentioned in Subchapter 2.1.4, which can be further discussed to complete the capacity research on urban mixed traffic.

Based on the results of road traffic influences in this dissertation work, it is also valuable to figure out the loss of urban rail-bound transport capacity and extra time caused on urban rail-bound transport and even on road traffic in the economic view for further expansion. In the mixed traffic zone, the interactions among urban mixed traffic result in economic costs due to the capacity and time loses not only on urban rail-bound transport but also on road traffic. It is valuable to evaluate the investigated mixed traffic zone is road traffic friendly or much friendlier to urban rail-bound transport. Furthermore, it is meaningful to estimate the feasibility if possible measures (new tunnel or over bridge for example to separate the urban rail-bound transport from road traffic) can be adopted to replace the mixed traffic zone. Further research can be focused on the economic view that is integrated with the standardized evaluation for infrastructure investment of public transport of the possible measures.

In conclusion, the findings in this dissertation can expand the application of capacity research on various mixed traffic zones, in which the urban rail-bound transports are influenced by road traffic. It is an important step for further researches to complete the capacity research on rail-bound system.

Appendix I: Basic Information of the Investigated Example

An investigated example for capacity research of urban rail-bound transport with road traffic influences in a mixed traffic zone of level crossing in a whole investigated area, which is shown in Appendix Figure 1. It is a partial infrastructure of a real urban rail-bound transport network (Stadtbahnnetz) with a mixed traffic zone of level crossing shown in the red circle. There are 25 scheduled stops in the whole investigated area.

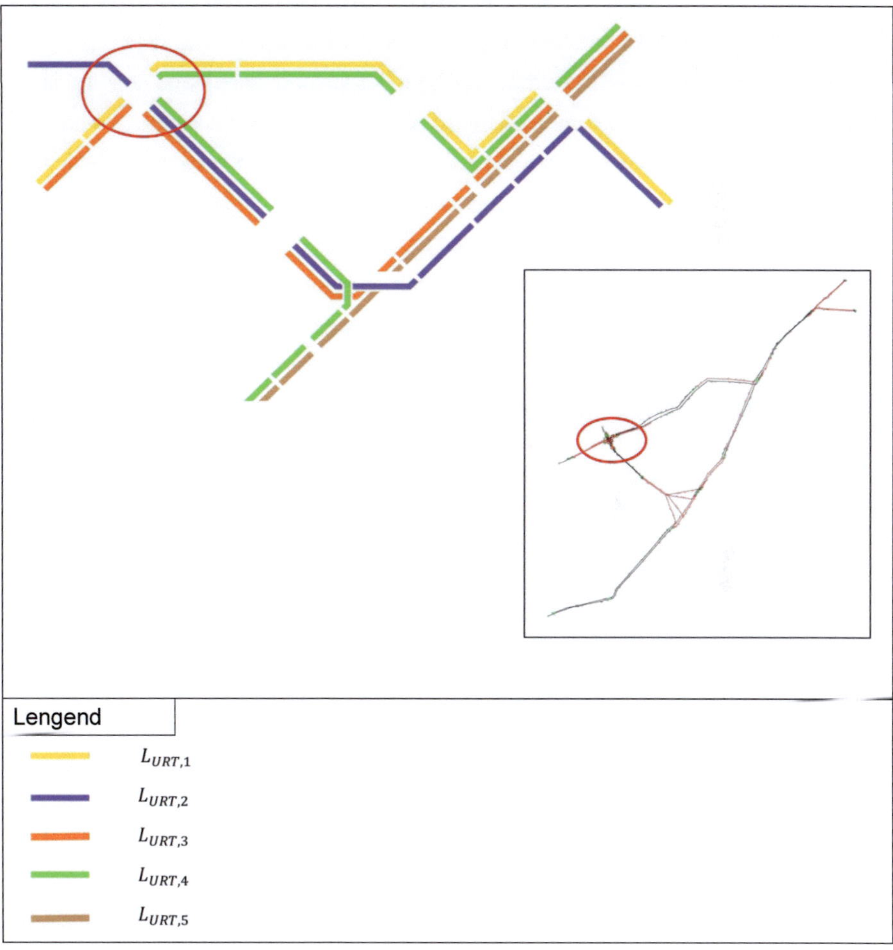

Appendix Figure 1: The Infrastructure of the Investigated Example with a Mixed Traffic Zone of Level Crossing

There are 5 lines of urban rail-bound transports in the whole investigated area and four of them pass through the mixed traffic zone of level crossing (as the topology structure of rail tracks for urban rail-bound transport in Figure 3-1). The basic operating program is shown in Appendix Table 1. The five lines of urban rail-bound transports are all in bidirectional operations with 6 trains per hour.

Four of them passing through the level crossing do experience the derived perturbation caused by road traffic for capacity research with distribution approach. For the modeling approach, the operating program should be built with the "trains" of urban rail-bound transports (i.e. the urban rail-bound transport lines) and the trains of "modeled road traffic" (i.e. road traffic lines at level crossing) for capacity research. Therefore, 180 trains per hour are scheduled in operation in the whole investigated area with 60 "trains" as urban rail-bound transport, 48 trains running through the level crossing, and 120 trains of "modeled road traffic". For the modeling of event-driven system with the developed algorithm, the four lines of urban rail-bound transports running through the mixed traffic zone of level crossing are modeled with semi-random arriving time based on the given operating program and stepwise-varied traffic flows to trigger the event for simulation.

Operating Program		
Train Mixture		**Traffic Flow for each direction [trains/h]**
Urban Rail-bound Transport Lines	$L_{URT,1}$	6
		6
	$L_{URT,2}$	6
		6
	$L_{URT,3}$	6
		6
	$L_{URT,4}$	6
		6
	$L_{URT,5}$	6
		6
Road Traffic Lines at the Level Crossing	$L_{RT,1}$	30
	$L_{RT,2}$	30
	$L_{RT,3}$	30
	$L_{RT,4}$	30

Appendix Table 1: The Corresponding Operating Program of the Investigated Example

Appendix II: Results of Capacity Research for the Investigated Example

For the investigate example described in Appendix I, there are about 280 stochastic timetables with stepwise-varied traffic flows ranging from 5% to 250% with each step at 5% created to determine the throughput capacity and waiting time function with various approaches, and further derivation of the recommended area of traffic flow for evaluation of operating performance. There are four scenarios:

a) without road traffic influences, which can be treated as the railway system
b) with road traffic influences, using the modeling approach
c) with road traffic influences, using the distribution approach
d) with road traffic influences, using the model of event-driven system based on the developed algorithm

Throughput Capacity

Appendix Figure 2 shows the results of throughput capacity of the first three scenarios for the whole investigated mixed traffic zone of level crossing based on the method by [Schmidt 2009] and [Chu 2014] through the exit traffic flow and entry traffic flow. It is obvious that the throughput capacity without road traffic influences of the investigated area is 146.3 trains/h and much higher than that with road traffic influences at the level crossing. With the modeling approach, the derived throughput capacity is 103.1 trains/h, which is nearly the same as the result derived from the distribution approach at 104.3 trains/h.

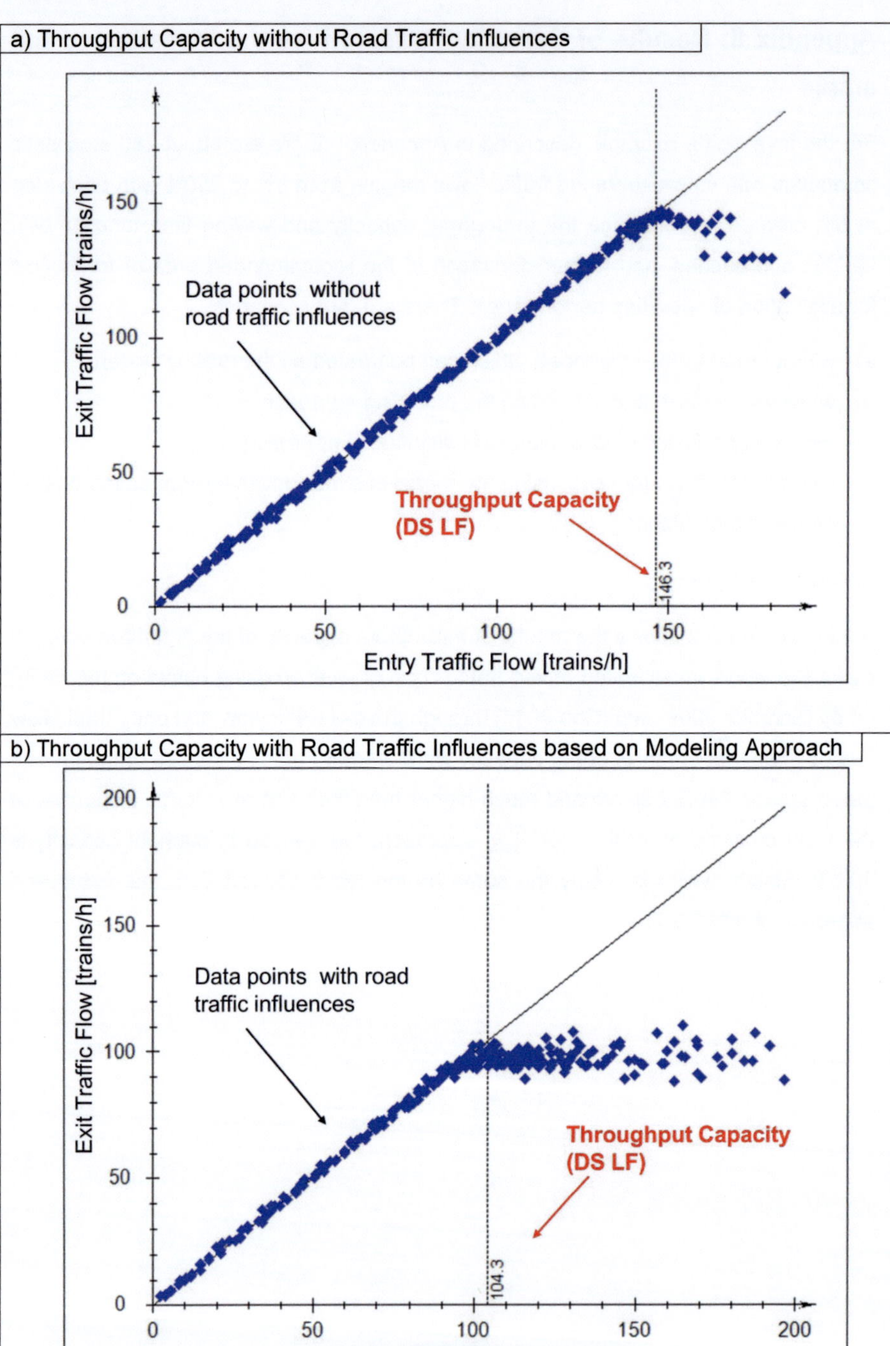

a) Throughput Capacity without Road Traffic Influences
Exit Traffic Flow [trains/h]
150
100
50
0
Data points without road traffic influences
Throughput Capacity (DS LF)
146.3
0
50
100
150
Entry Traffic Flow [trains/h]

b) Throughput Capacity with Road Traffic Influences based on Modeling Approach
Exit Traffic Flow [trains/h]
200
150
100
50
0
Data points with road traffic influences
Throughput Capacity (DS LF)
104.3
0
50
100
150
200
Entry Traffic Flow [trains/h]

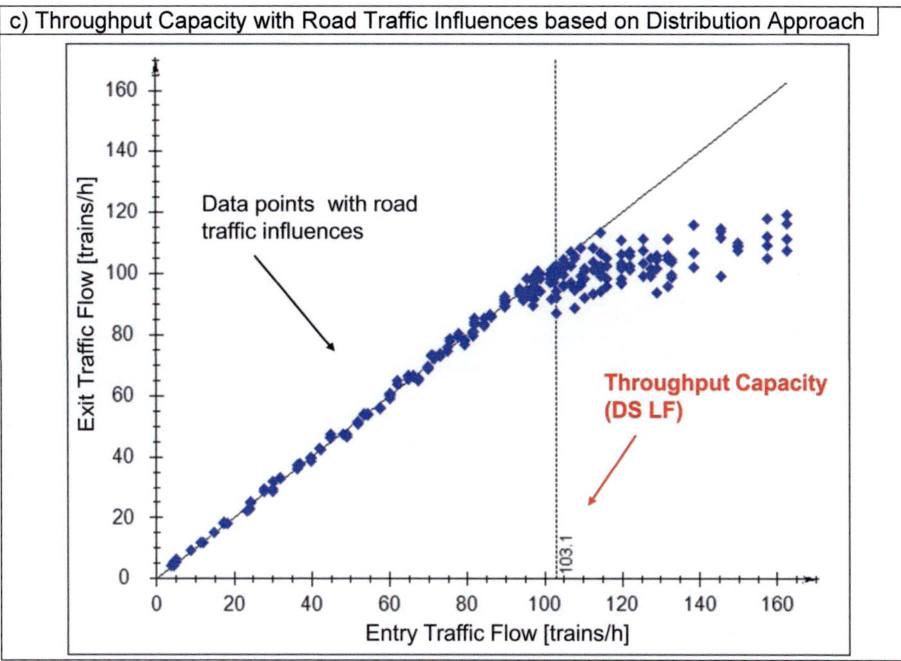

Appendix Figure 2: Throughput Capacity with and without Road Traffic Influences

The relationship between potential occupation time for road traffic (O_{RT}) and the hindrance time of urban rail-bound transport by road traffic (H_{URT}) can derive the approximate throughput capacity as described in Subchapter 5.5.1. As shown in Appendix Figure 3, there are some fluctuations along both curves, however, the overall trends of the two indicators with increasing traffic flow of urban rail-bound transport is to reach the stable and approach each other. In this example, the evaluated investigated time period is set with the 50 traffic cycles in the middle of the investigated time period of 150 traffic cycles in the investigated level crossing.

Consequently, the throughput capacity is approximately determined at the stable point with the traffic flow of 82 trains/h. In addition, the newly added term with the parameter d in the adapted waiting time function (4-11) is determined.

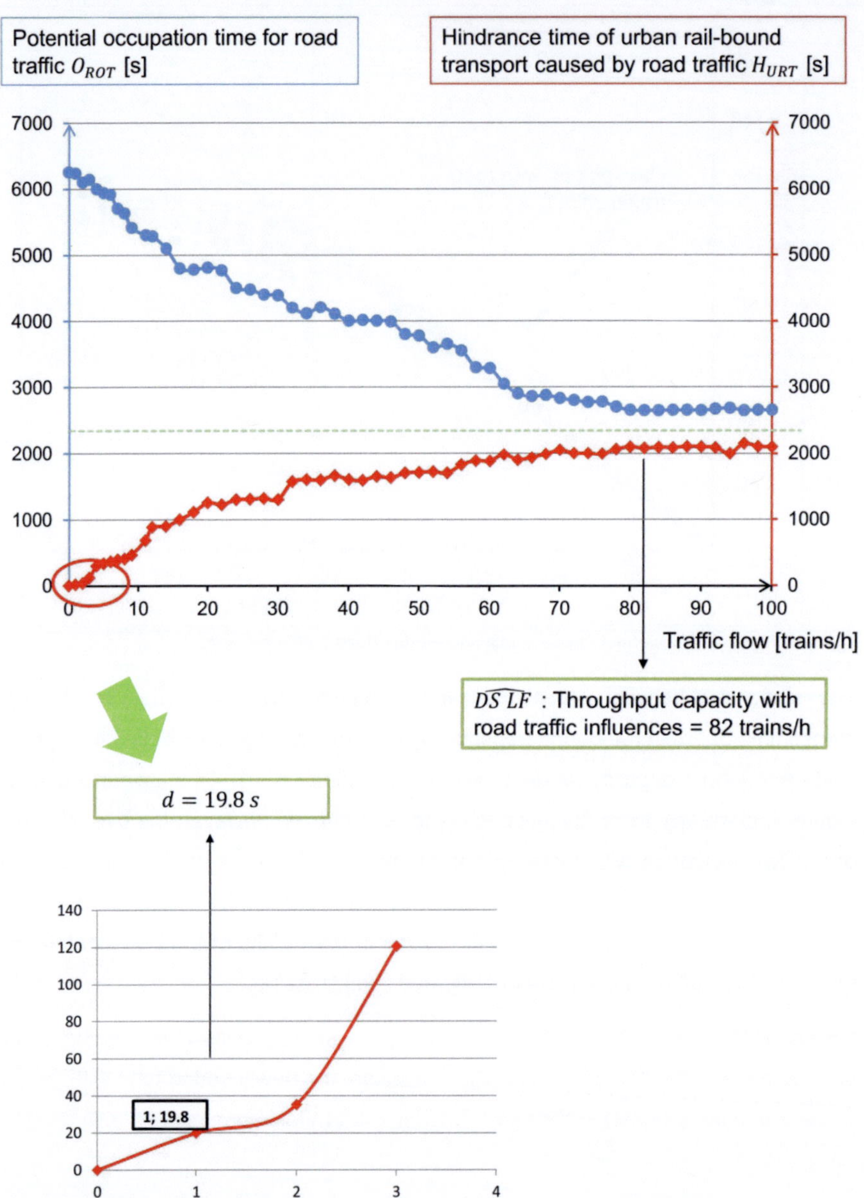

Appendix Figure 3: Throughput Capacity of Scenario d) for the Investigated Mixed Traffic Zone of Level Crossing with the Event-driven System

Waiting time function

As shown in Appendix Table 2, Figure 3-4 and Figure 3-7, the waiting time function (4-11) can fit the data points of the scenarios with road traffic influences better than the existing one (4-7) that is more suitable for the scenario of railway system without road traffic influences.

		Waiting Time Function		Coefficient of Determination	Adjusted Coefficient of Determination
Without Road Traffic Influences		$ET_w = a_2 \cdot \dfrac{\eta}{(1-\eta)^{b_2}} \cdot (1 - \eta^{c_2})$	(4-7)	0.9918	0.9917
With Road Traffic Influences	Modeling Approach	$ET_w = a_2 \cdot \dfrac{\eta}{(1-\eta)^{b_2}} \cdot (1 - \eta^{c_2})$	(4-7)	0.9848	0.9847
		$ET_w = \dfrac{a_5 \cdot \eta \cdot (1 - \eta^{c_5})}{(1-\eta)^{b_5}} + \dfrac{d_5}{(1-\eta)^{2/3}}$	(4-11)	0.9928	0.9927
	Distribution Approach	$ET_w = a_2 \cdot \dfrac{\eta}{(1-\eta)^{b_2}} \cdot (1 - \eta^{c_2})$	(4-7)	0.9831	0.9829
		$ET_w = \dfrac{a_5 \cdot \eta \cdot (1 - \eta^{c_5})}{(1-\eta)^{b_5}} + \dfrac{d_5}{(1-\eta)^{2/3}}$	(4-11)	0.9922	0.9921

Appendix Table 2: (Adjusted) Coefficient of Determination of Waiting Time Function for Various Scenarios

Recommended area of traffic flow

The derived recommended area of traffic flow is based on the waiting time function (4-11) for the scenarios with road traffic influences. The result with the modeling approach with the upper limit of 91.7 trains/h is almost the same to the result with distribution approach at 92.2 trains/h. They are definitely lower than the scenario without road traffic influences at 139.8 trains/h. Similar to the lower limit of the recommended area of traffic flow, this big drop is mainly due to the road traffic influences at the level crossing in the whole investigated area.

Scenarios		Waiting Time Function		Recommended Area of Traffic Flow
Without Road Traffic Influences		$ET_w = a_2 \cdot \dfrac{\eta}{(1-\eta)^{b_2}} \cdot (1 - \eta^{c_2})$	(4-7)	93.9 – 139.8 trains/h
With Road Traffic Influences	Modeling Approach	$ET_w = \dfrac{a_5 \cdot \eta \cdot (1 - \eta^{c_5})}{(1-\eta)^{b_5}} + \dfrac{d_5}{(1-\eta)^{2/3}}$	(4-11)	60.3 – 91.7 trains/h
	Distribution Approach	$ET_w = \dfrac{a_5 \cdot \eta \cdot (1 - \eta^{c_5})}{(1-\eta)^{b_5}} + \dfrac{d_5}{(1-\eta)^{2/3}}$	(4-11)	63.6 – 92.2 trains/h

Appendix Table 3: Recommended Area of Traffic Flow based on Corresponding Waiting Time Function for Various Scenarios

Appendix III: Supplementary Logics of Execution of Cases for Level Crossing

Case 1:

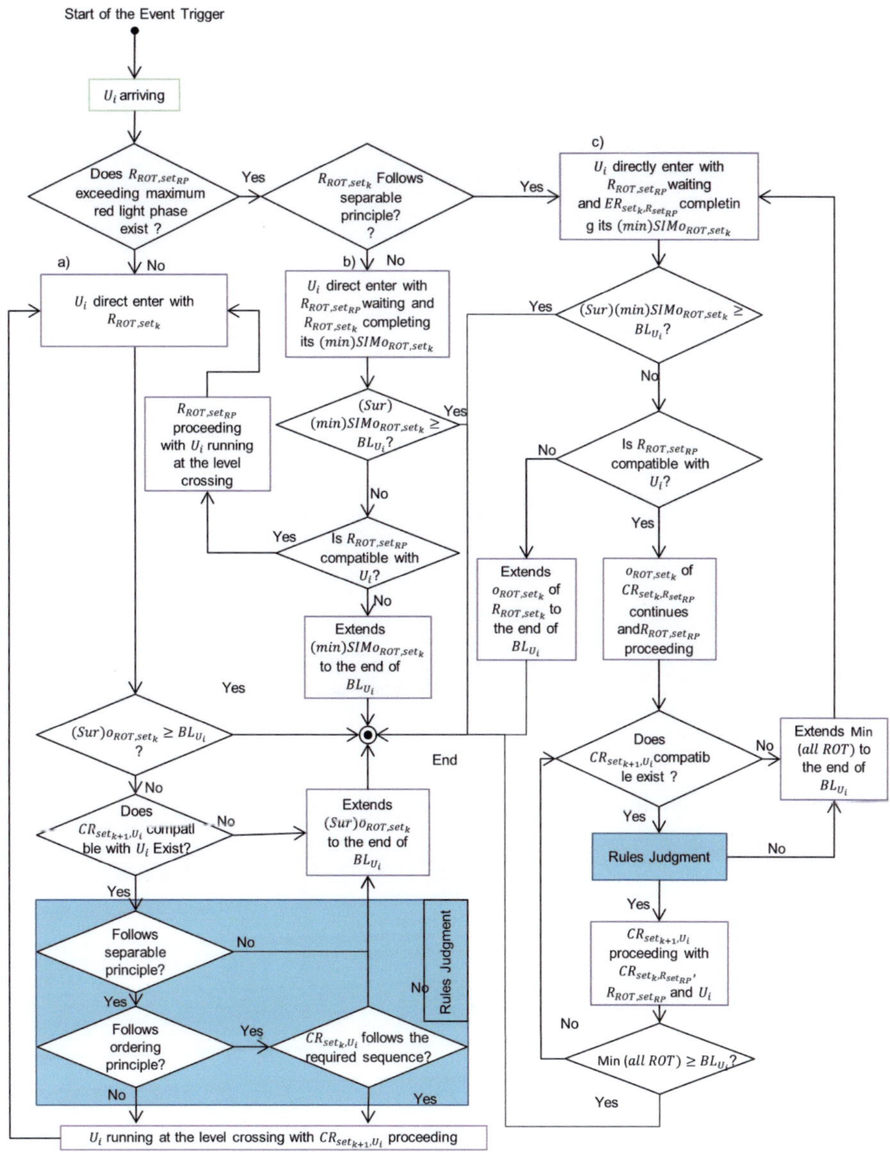

Appendix Figure 4: Workflow of Execution of *LC* Cases 1

Case 2:

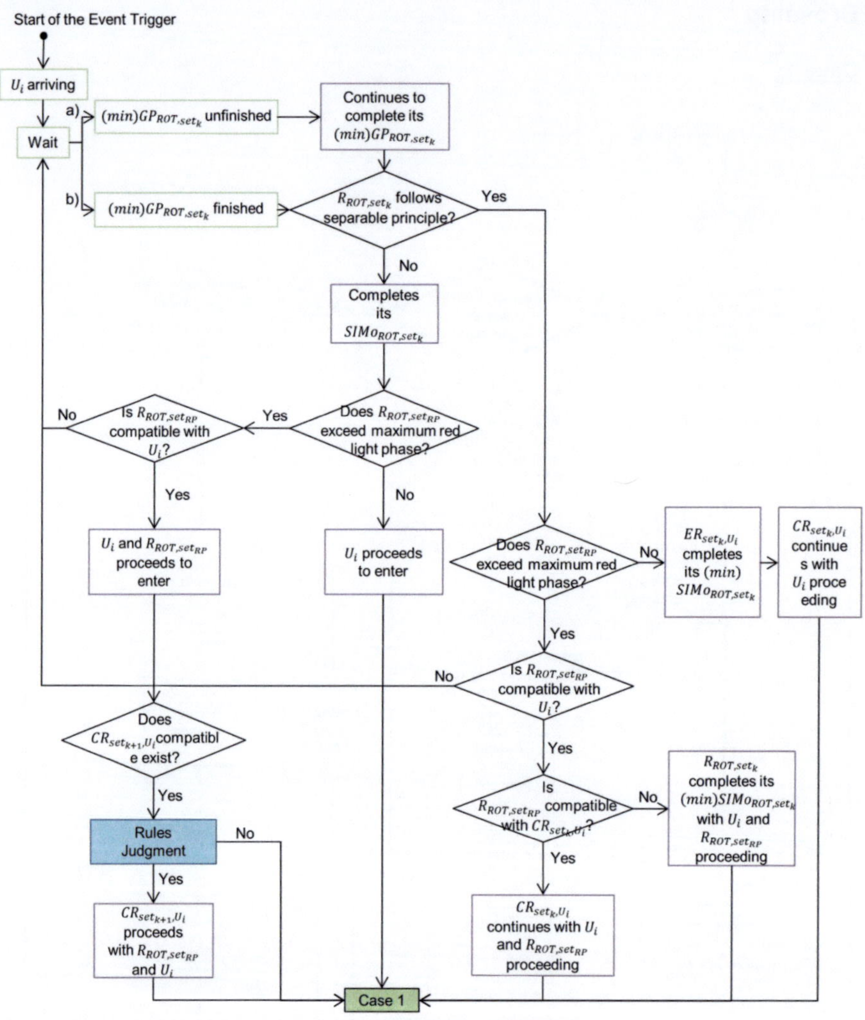

Appendix Figure 5: Workflow of Execution of LC Cases 2

Case 3:

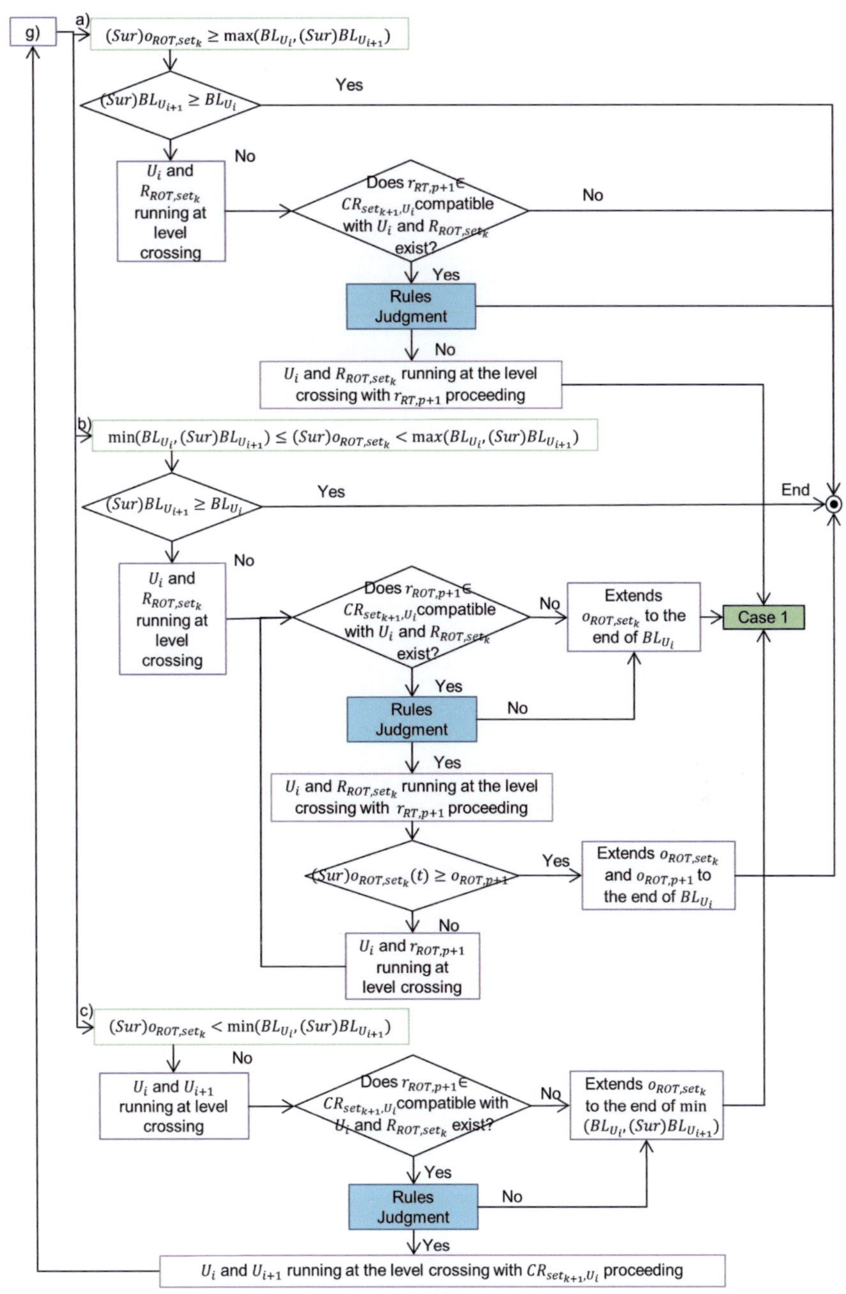

Appendix Figure 6: The Workflow of Execution of the Probability g) in *LC* Case 3

Case 4:

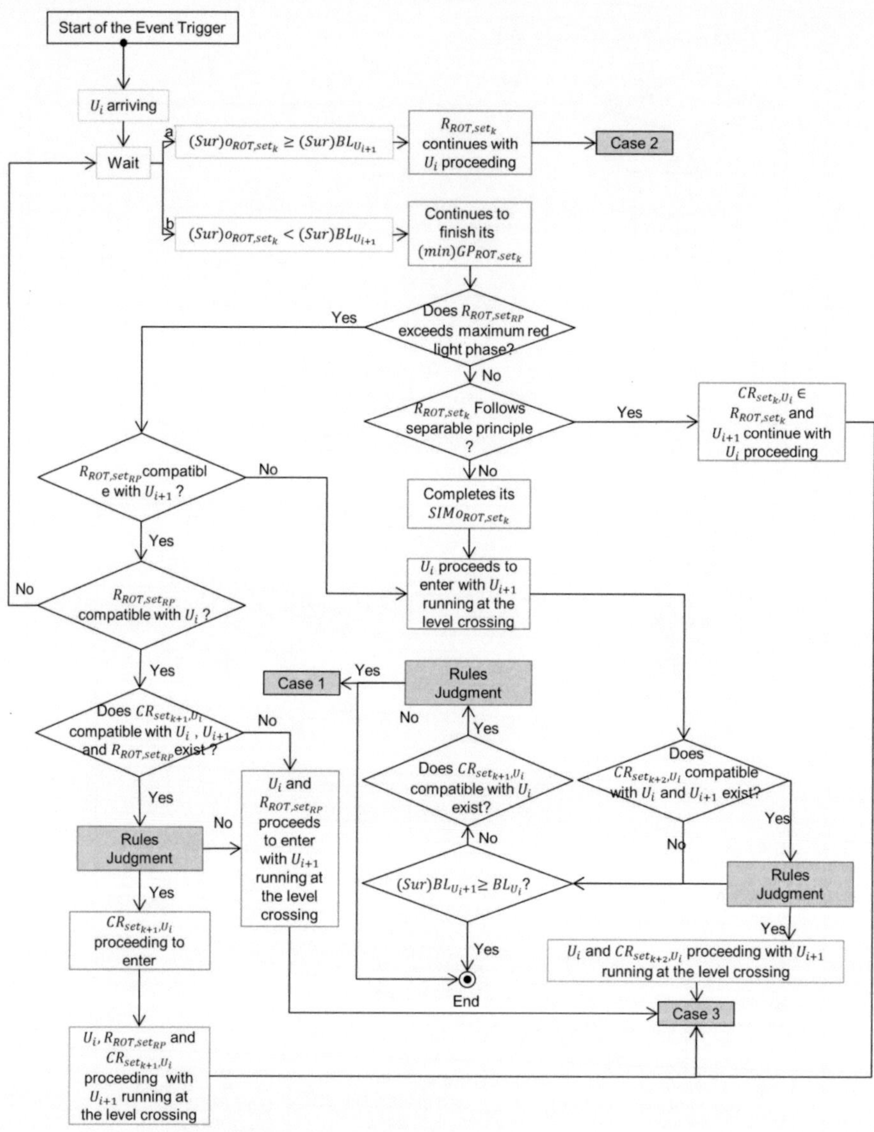

Appendix Figure 7: Workflow of Execution of LC Cases 4

List of Abbreviations

East	E/E_1
Location Point	LP
Level Crossing	LC
Mixed Traffic Zone	MTZ
North	N/N'
Programm zur Untersuchung des Leistungsverhaltens von Eisenbahninfrastrukturen	PULEIV
Recovery Time	REC
Road Traffic	ROT
South	S/S'
Shared Road	SR
Shared Space	SS
Urban Rail-bound Transport	URT
Verkehr In Städten – SIMulationsmodell	VISSIM
West	W/W_1

Formula Symbols

a, b, a_i, b_i, c_i, d_i	Parameters of waiting time function
α	The nonlinear parameters of the model function $\hat{y}_i(x)$
β	The linear parameters of the model function $\hat{y}_i(x)$
$B_{bef}(\eta)$	Traffic energy function of the waiting time
$deg.$	The degrees of freedom of the waiting time function
del_m	Average value of delays in minute
del_{max}	Maximum delay for perturbed urban rail-bound transport in minute
ϕ_j	The j th of n basis function of x
EL_v	Average number of requests in queuing system
ET_a	Expected value of time interval between arrival times of two trains
ET_b	Expected value of the service time in the server of queuing system
$ETw(\eta)$	(statistical) Expected value waiting time function
η	Utility Factor of Capacity
λ	Average arrival frequency (rate)
μ	Average service rate of a single server
ρ	Utility factor
$GP_{ROT,p}$	The green light phase of road traffic route $r_{ROT,p}$
h	The number of times that the urban rail-bound transport is hindered along the shared road
hl_i	The leverages that adjust the residuals by reducing the weight of high leverage data points

$h_{URT,p}$	The hindrance time of urban rail-bound transport route caused by road traffic route $r_{ROT,p}$ in one traffic cycle at an investigated mixed traffic zone
$h_{URT,q,p}$	The effective hindrance time of an urban rail-bound transport route $r_{URT,q}$ caused by road traffic route $r_{ROT,p}$
H_{URT}	The hindrance time of urban rail-bound transport caused by road traffic
K	A tuning constant equal to 4.685
l	The distance from the location point (LP_l), at which the urban rail-bound transport is hindered by road traffic firstly to the end point of the shared road
LP_l	Defined location point along the shared road with the index l
L_{ROT}	The lines of rail tracks for trains of "modeled road traffic"
L_{URT}	The lines of rail tracks for "trains" of urban rail-bound transports
m	The number of the road traffic routes with $|R_{ROT}| = m$
MAD	The median absolute deviation of the residuals.
$MS(Res)$	The residual mean of squares
$MS(Total)$	The total mean of squares
MTZ_i	The mixed traffic zone in a whole investigated area with the index i
n	The number of the urban rail-bound transport routes with $|R_{URT}| = n$
n_{req}	Possible number of requests in queuing system
n_s	The number of sample (data points)
N	Traffic flow (trains/h)

Formula Symbol

N_{req}	The number of requests in queuing system
$o_{RT,p}$	the effective occupation time of the road traffic route $r_{ROT,p}$
O_{ROT}	The potential occupation time for road traffic
$pro.$	Proportion values of perturbed urban rail-bound transport (delayed urban rail-bound transport) in %
P_{ROT}	The perturbation parameters for perturbation caused by road traffic influences
P_{URT}	The perturbation parameters for perturbation without road traffic influences
r_{adj}	the usual least squares residuals
res_i	The residual for the i th data point
R^2	Coefficient of determination
$\bar{R}^2$	Adjusted coefficient of determination
$r_{ROT,p}$	The road traffic route with the index $p \in [1, m]$
R_{ROT}	The set of routes of road traffic
$r_{URT,q}$	The urban rail-bound transport route, with the index $q \in [1, n]$
R_{URT}	The set of routes of urban rail-bound transport
$RP_{ROT,p}$	The red light phase of road traffic route $r_{ROT,p}$
S	The length of the shared road
s_i	Traffic control signal with number of i
STF	The traffic flow of all urban rail-bound transports in the scheduled timetable
$S_{W,ROT/URT}$, $S_{E,ROT/URT}$, $S_{N,ROT/URT}$ and $S_{S,ROT/U1}$	The stops for mixed traffic with directions of the investigated example

S_{rob}	The robust variance given by $MAD/0{,}6745$
$S_{rel}(\eta)$	Relative sensitivity of the waiting time function
SS	The summed square of residuals where m the number of data points included in the fit is
$SS(Res)$	The residual sum of squares
$SS(Total)$	The total sum of squares
t	The index of the investigated time period, $t \in [1,\ T]$
T	Investigated time period in hour
$DS\ LF$	Throughput capacity
$\widehat{DS\ LF}$	Throughput capacity with road traffic influences
T_{sl}	The time slice in second
u_i	The standardized adjusted residuals of i th iterative
U_i	A train of urban rail-bound transport on its corresponding route $r_{URT,q+1}$ with the index i
v	The velocity of urban mixed traffic (road traffic)
v_{bef}	Average transport velocity
w_i:	the robust bisquare weights $w_{bisqure}$ of i th iterative
x_i	The variables of the model function $\hat{y}_i(x)$
y_i	The i th of m fitted value with model function
$\hat{y}_i$	The i th of m fitted value with model function
$YP(G)_{ROT,p}$	The yellow light phase of road traffic $r_{ROT,p}$ changed from its green light phase
$YP(R)_{ROT,p}$	The yellow light phase of road traffic $r_{ROT,p}$ changed from its red light phase

Glossary

Cycle time

Cycle time is the time duration that it takes to complete one traffic signal cycle at a level crossing.

Effective occupation time of road traffic route

Effective occupation time of road traffic route is the time duration that it takes to complete one road traffic route ($r_{ROT,p}$) starting from its entry until another exclusive mixed traffic route (road traffic or urban rail-bound transport route) is allowed to enter the mixed traffic zone.

Effective hindrance time of urban rail-bound transport route

The effective hindrance time of an urban rail-bound transport route is the time duration that an urban rail-bound vehicle is hindered by the occupation of road traffic at an investigated mixed traffic zone.

Hindrance time of urban rail-bound transport caused by road traffic

The hindrance time of urban rail-bound transport caused by road traffic is the total occupation time of the road traffic, at which the urban rail-bound vehicles are hindered by the occupation of the road traffic. It is a time interval that is the maximum waiting time of one or more hindered urban rail-bound vehicles in the mixed traffic zone that is occupied by one road traffic route.

Instantaneous values of indicators

Instantaneous values are the values of indicators (del_m and $pro.$) that are the simulation results of output delays related to the given perturbation parameters.

Level crossing

A level crossing is an intersection where an urban rail-bound line crosses an urban roadway or path at the same level. The term also applies when an urban rail-bound line with separate right-of-way or reserved track crosses a road in the same manner.

Modeled road traffic

Modeled road traffic are the trains representing the road traffic at the investigated mixed traffic zone modeled in the simulation tool RailSys for the modeling approach.

Perturbation

Perturbations are the disruptions to the scheduled timetable during operations. They are introduced into the simulation model on rail-bound transport in the scheduled timetable.

Perturbation parameters

Perturbation parameters can be used to describe the distribution of one perturbation in the simulation model. There are three perturbation parameters: average delay, proportion values of perturbed urban rail-bound transport, and maximum delay for delayed urban rail-bound transport.

Perturbation type

There are three types of perturbations studied in this research project: initial delay, dwell time extension, departure time extension and running time extension.

Potential occupation time for road traffic O_{ROT}

Potential occupation time for road traffic is the maximal possible occupation time of the road traffic at an investigated mixed traffic zone in an investigated time period. It is a time interval that is maximally allocated to the exclusive use of road traffic route at an investigated mixed traffic zone in an investigated time period.

Recovery time

Recovery time is a time supplement that is added to the pure running time and dwell time to enable a train to make up small delays [Pachl 2014].

Road traffic

Road traffic refers to all traffic types moving on the

urban roadway, which can be further categorized into motorized road traffic and non-motorized road traffic. The motorized road traffic includes the private vehicles, like truck, private car, motor (-bicycle); and public vehicles, like bus. The non-motorized road traffic includes pedestrians and bicycles.

Route

Route is defined as a section of infrastructure for urban rail-bound transport or an urban roadway for road traffic with a specific direction, on which the mixed traffic can move in the investigated mixed traffic zone from the entry point A to exit point B.

Shared road

Shared road is one basic type of mixed traffic zone, where the rail tracks for urban rail-bound transport share with the urban roadways and even the bikeways (bicycle lanes) for road traffic.

Target values of indicators

Target values of indicators are the values of indicators (del_m and $pro.$) that are given from collected operational data in reality as reference value for calibrating.

Traffic light phase

Traffic light phase is defined as the time duration to direct the movements of road traffic controlled by the corresponding traffic control signals. It can give the indication for movements of road traffic with certain time duration. There are four traffic light phases discussed in this research project, the red light phase, the green light phase, and two yellow light phases changed from either the red light or the green light.

Traffic control signal

Traffic control signal is also known as a traffic light, which is a signaling device positioned at intersections, pedestrian crossings and other locations to control the behaviors of traffic loads, whether to move or be still with corresponding indications of the traffic light.

Urban mixed traffic

The different modes of transports that operate in urban mixed traffic zones including urban rail-bound transport are collectively called urban mixed traffic. It can be categorized into public transport (urban rail-bound transport and bus) and private transport (individual vehicle and pedestrian) or into urban rail-bound transport and road traffic that is mainly used in this research project.

Urban mixed traffic zone

Urban mixed traffic zone is an urban area, in which not only urban rail-bound transport but also mutual interfered other transports (urban mixed traffic) operate.

Urban rail-bound transport

Urban rail-bound transport is the rail-bound system, which operates in urban area, including metro, light rail transit and tram etc..

Bibliography

[Blue et al. 1997]

Blue, V. J.; Embrechts, M. J.; Adler, J. L.: *Cellular Automata Modeling of Pedestrian Movements.* IEEE, 1997.

[Chu 2014]

Chu, Zifu: *Modellierung der Wartezeitfunktion bei Leistungsuntersuchungen im Schienenverkehr unter Berücksichtigung der transienten Phase.* Dissertation, Universität Stuttgart, Institut für Eisenbahn- und Verkehrswesen. Betreut von Ullrich Martin. Schriftenreihe VWI Neues verkehrswissenschaftliches Journal - Ausgabe 10. Norderstadt: BoD - Book on Demand., 2014.

[Cui et al. 2014]

Cui, Yong; Ullrich, Martin; Zhao, Weiting; Cao, Nan: *Entwicklung eines Algorithmus für die Kalibrierung von Modellen zur Betriebssimulation in spurgeführten Verkehrssystemen unter Berücksichtigung stochastischer Bedingungen.* Deutsche Forschungsgemeinschaft (DFG) - Forschungsvorhaben (GZ: MA 2326/9-1). Schriftenreihe VWI Neues verkehrswissenschaftliches Journal - Ausgabe 9. Norderstadt: BoD - Book on Demand., 2014.

[DB Netz AG 2008]

DB Netz AG: *DB Richtlinie 405 - Fahrwegkapazität,* 2008.

[Draper & Smith 1998]

Draper, Norman R.; Smith, Harry: *Applied regression analysis.* Wiley, New York, Chichester, 1998.

[Everitt 2002]

Everitt, Brian: *The Cambridge dictionary of statistics.* Cambridge University Press, Cambridge, 2002.

[Fellendorf & Vortisch 2010]

Fellendorf, Martin; Vortisch, Peter: *Microscopic Traffic Flow Simulator VISSIM.* In: Barceló, J. (Hrsg.): Fundamentals of Traffic Simulation. Springer New York, Editor: Barceló, Jaume, New York, NY, 2010, S. 63–93.

[Fukui & Ishibashi 1996]

Fukui, Minoru; Ishibashi, Yoshihiro: *Traffic Flow in 1D Cellular Automaton Model Including Cars Moving with High Speed.* Journal of the Physical Society of Japan, 65, 1996, 6, pp. 1868–1870.

[Green 1984]

Green, Linda: *A Queueing System with Auxiliary Servers*. Management Science, 30, 1984, pp. 1207–1216.

[Hertel et al. 1987]

Hertel, Günter; Ludwig, Dietmar; Bauer, Jörg: *Leistung und Qualität im Eisenbahntransport*. In: Wissenschaftliche Zeitschrift Hochschule für Verkehrswesen Dresden, Nr. 2, Jg. 34, 1987, pp. 207–234.

[Hertel 1992]

Hertel, Günter: *Die maximale Verkehrsleistung und die minimale Fahrplanempfindlichkeit auf Eisenbahnstrecken*. In: ETR-Eisenbahntechnische Rundschau, Nr. 10, Jg. 41., 1992, pp. 665–671.

[International Union of Railways (UIC) 2004]

International Union of Railways (UIC): *UIC CODE 406*: Capaciy, 2004.

[Janecek et al. 2010b]

Janecek, David; Weymann, Frédéric Herbert Georges; Schaer, Th.: *LUKS - integriertes Werkzeug zur Leistungsuntersuchung von Eisenbahnknoten und -strecken*. In: ETR-Eisenbahntechnische Rundschau, 59 (1+2), 2010b, pp. 25–32.

[Janecek & Weymann 2010a]

Janecek, David; Weymann, Frédéric Herbert Georges: *LUKS - Analysis of lines and junctions*. In: 12th World Conference on Transport Research, Instituto Superior Tecnico, Lisboa, 2010a.

[Kontaxi & Ricci 2010]

Kontaxi, E.; Ricci, S.: *Capacity Analysis: Methodological Framework and Harmonization Perspectives*, University of Rome, DICEA Department, Italy. 12th WCTR. Lisbon, Portugal, 2010.

[Li 2015]

Li, Xiaojun: *Mikroskopische Engpassanalyse bei eisenbahnbetriebswissenschaftlichen Leistungsuntersuchungen*. Dissertation, Universität Stuttgart, Institut für Eisenbahn- und Verkehrswesen. Betreut von Ullrich Martin. Schriftenreihe VWI Neues verkehrswissenschaftliches Journal - Ausgabe 14. Norderstadt: BoD - Book on Demand., 2015.

[Lindner 2011]

Lindner, Tobias: *Applicability of the Analytical Compression Method for Evaluating*

Node Capacity. Universität Braunschweig, Institut für Eisenbahnwesen und Verkehrssicherung IfEV, 2011.

[Ludwig 1990]

Ludwig, Dietmar: *Beitrag zur Leistungs- und Qualitätssicherung von Streckenfahrplänen der Eisenbahn*. Dissertation, Dresden, 1990.

[Martin et al. 2011]

Martin, Ullrich; Schmidt, Christine; Chu, Zifu: *PULEIV Anwenderleitfaden. zur PULEIV - Version 2.1. Anleitung zur Ermittlung des Leistungsverhaltens von Eisenbahninfrastrukturen mit Hilfe von Simulationsprogrammen.*, Stuttgart, 2011.

[Martin et al. 2012a]

Martin, Ullrich; Li, Xiaojun; Warninghoff, Carsten-Rainer: *Bewertungsverfahren für Knotenelemente bei der Infrastrukturbemessung - Replan.* In: ETR-Eisenbahntechnische Rundschau, Nr. 11, Jg. 61, 2012, pp. 38–43.

[Martin et al. 2012b]

Martin, Ullrich; Chu, Zifu; Peter Breuer: *Leistungsuntersuchung Staatsgalerie im Rahmen der Umbaumaßnahmen zu Stuttgart 21.* Abschlussbericht (unpublished), Im Auftrag der Stuttgarter Straßenbahnen AG, 2012.

[Martin 2014]

Martin, U. (Hrsg.): *Railway Timetabling & Operations*: 12. Performance Evaluation. Chapter 9: Simulation. Eurailpress in DVV Media Group. Editors: Hansen, Ingo A.; Pachl, Jörn, 2014.

[Martin & Chu 2012]

Martin, Ullrich; Chu, Zifu: *Dynamisierung von Zeitscheiben in Betriebsprogrammen bei Leistungsuntersuchungen.* In: ETR-Eisenbahntechnische Rundschau, Nr. 5, 2012, pp. 40–45.

[Martin & Chu 2013]

Martin, Ullrich; Chu, Zifu: *Direkte Experimentelle Bestimmung der Maximalen Leistungsfähigkeit bei Leistungsuntersuchungen im spurgeführten Verkehr.* Deutsche Forschungsgemeinschaft (DFG) - Forschungsvorhaben (GZ: MA 2326/6-1). Schriftenreihe VWI Neues verkehrswissenschaftliches Journal-NVJ, Ausgabe 7, Norderstadt: BoD - Books on Demand, 2013.

[Martin & Liu 2016a]

Martin, Ullrich; Liu, Di: *Capacity Research in Urban Rail-bound Transportation with

Special Consideration of Mixed Traffic. Deutsche Forschungsgemeinschaft (DFG) - Forschungsvorhaben (GZ: MA 2326/13-1). Schriftenreihe VWI Neues verkehrswissenschaftliches Journal - Ausgabe x. Norderstadt: BoD - Book on Demand., 2016.

[Martin & Liu 2016b]

Martin, Ullrich; Liu, Di: *Einfluss des Straßenverkehrs auf die Wartezeitfunktion bei Leistungsuntersuchungen im SPNV*. In: ETR-Eisenbahntechnische Rundschau, Nr. 1+2, 2016, pp. 21–25.

[Michl & Sojka 2014]

Michl, Zdeněk; Sojka, Martin: *Microsimulation as a tool for evaluation of infrastructure and operational concept alternatives in a complex railway node*. in: Logistyka, 4, 2014, pp. 3029–3037.

[Moler 2008]

Moler, Cleve B.: *Numerical computing with MATLAB*. Society for Industrial and Applied Mathematics (SIAM), Philadelphia, ©2004, 2008.

[Nagel & Schreckenberg 1992]

Nagel, Kai; Schreckenberg, Michael: *A Cellular Automaton Model for Freeway Traffic*. Journal de Physique I, Vol. 2 (12), 1992, pp. 2221–2229.

[Newell 2002]

Newell, G. F.: *A simplified car-following theory: A lower order model*. Transportation Research Part B: Methodological, 36, 2002, pp. 195–205.

[Nießen 2008]

Nießen, Nils: *Leistungskenngrößen für Gesamtfahrstraßenknoten*. Dissertation, Aachen, 2008.

[Omahen 1977]

Omahen, Kenneth J.: *Capacity Bounds for Multiresource Queues*. J. ACM (Journal of the Association for Computing Machinery), 24, 1977, pp. 646–663.

[Pachl 2014]

Pachl, Jörn: *Railway Operation and Control*. VTD Rail Publishing, Mountlake Terrace (USA), 2014.

[Pachl 2016]

Pachl, Jörn: *Systemtechnik Des Schienenverkehrs*: Bahnbetrieb Planen, Steuern Und Sichern. Vieweg + Teubner Verlag, 2016.

[Potthoff 1972]

Potthoff, Gerhardt: *Verkehrsströmungslehre*. VEB Verlag für Verkehrswesen Berlin, 1972.

[RMCON 2010]

RMCON: *Manual/ Handbuch RailSys 7*, 2010.

[Rossetti 2009]

Rossetti, Manuel D.: *A Simulation-based Approach for Estimating the Commercial Capacity Railways*, IEEE, 2009.

[Schmidt 2009]

Schmidt, Christine: *Beitrag zur experimentellen Bestimmung der Wartezeitfunktion bei Leistungsuntersuchungen im spurgeführten Verkehr*. Dissertation, Universität Stuttgart, Institut für Eisenbahn- und Verkehrswesen, 2009.

[Siefer 2014]

Siefer, T. (Hrsg.): *Railway Timetabling & Operations*: 9. Simulation. Eurailpress in DVV Media Group. Editors: Hansen, Ingo A.; Pachl, Jörn, 2014.

[Straßenverkehrs - Ordnung 2015]

Straßenverkehrs - Ordnung (StVO): *Allgemeine Verwaltungsvorschrift Allgemeine Verwaltungsvorschrift zur Straßenverkehrs-Ordnung (VwV-StVO)*, Bundesministerium für Verkehr, Bau und Stadtentwicklung (BMVBS),Bundesregierung, Deutschland, 2015.